THE UNITED STATES
AND WORLD
ENERGY SOURCES

THE UNITED STATES AND WORLD ENERGY SOURCES

edited by

Larry L. Berg
Lawrence M. Baird
Emilio E. Varanini III

PRAEGER

PRAEGER SPECIAL STUDIES • PRAEGER SCIENTIFIC

Library of Congress Cataloging in Publication Data

Main entry under title:

The United States and world energy sources.

1. Power resources. 2. Energy policy—United
States. I. Berg, Larry L. II. Baird, Lawrence.
III. Varanini, Emilio, E.
TJ163.2.U55   338.8′232   81-21171
ISBN 0-03-059807-9   AACR2

Published in 1982 by Praeger Publishers
CBS Educational and Professional Publishing
a Division of CBS Inc.
521 Fifth Avenue, New York, New York 10175 U.S.A.

23456789   145   987654321

Printed in the United States of America

PREFACE

The editors want to thank Rosalind K. Loring, Dean of the College of Continuing Education, for her support and encouragement for this project from its formative time to the conference at the Davidson Center and ultimately this volume. We also want to thank the members and staff of the California Energy Commission for their support and for having the foresight to sponsor and assist research, conferences, and publication of the papers presented at the U.S.C. hosted conference, "California & World Oil: The Strategic Horizon." We hope the insights on the complexities of current international oil policies that are presented by the authors will contribute to the efforts of policymakers and the public to meet the challenges that will face the nation throughout the remainder of this century.

Dr. Baird also would like to express special thanks to Peggy Dole and Mark Zierling of the California Energy Commission staff for their assistance. Dr. Berg wishes to thank all of his staff at the U.S.C. Institute of Politics and Government for their invaluable assistance.

Finally, we want to thank our wives, Mary, Maggie, and Lee and our children for once again putting up with us on yet another time-consuming project.

L. Berg
L. Baird
E. Varanini

v

CONTENTS

LIST OF TABLES AND FIGURES

INTRODUCTION

      The United States will be dependent on Persian Gulf oil at least until the year 2000. Many energy analysts denied the political and economic significance of this dependence dilemma in the decade of the 1970s by simply arguing that the United States could regain energy independence with strong government support for the development of any one of a number of alternative energy sources. The proposed energy sources were extremely diverse in nature and ranged from increased domestic oil exploration and production to rapid development of synthetic fuels to construction of additional nuclear power plants to the development of alternative energy sources such as wind, biomass, and geothermal to increased solar applications to eventually the conservation of energy. As we proceed into the decade of the 1980s, another equally disturbing realization confronts energy policymakers. Not only is the United States dependent on foreign oil, but none of the proposed alternative energy sources, developed independently or in concert with others, can attain the desired energy independence now or in the next decade.

      This volume examines the economic, political, and social implications of the oil dependence dilemma facing the United States. Most of the contributors are energy consultants in the public or private sector. Their analyses of the changing oil situation and its impact on other energy policies reflect either an international, national, or regional perspective with a unique combination of pragmatic insights and academic analyses of these complex issues. While examining the various aspects of the energy dependence dilemma presented here, one critical theme will probably recur to the reader. That is, given the inadequate nature of the U.S. response to the 1973 and 1979 shortfalls in foreign oil supplies, how will we manage the projected future shortages in foreign oil supplies? Or transcending the management of another short-term oil crisis, the reader might address the same problem from a broader perspective, that is, how can the United States reduce its vulnerability to foreign oil shortages over the next 20 years?

Whether the reader is concerned about the more immediate or long-term energy problems that confront the United States, this book sets the stage for understanding how closely tied we have become to the other nations of the world as we compete and cooperate to develop and consume scarce energy resources. It also helps to explain how the solutions developed by policymakers in the United States to cope with energy problems impact domestically and on other nations. The various contributors provide a world outlook on energy demand and supply and also explain the difficulties in making such an assessment. They forecast Middle East oil supplies to the United States with uncertainty because of the political instability of many nations in the region. They also provide an assessment of critical supplies available to the United States from Mexico, Venezuela, and Indonesia and project supply availability from North American sources. Finally, and perhaps most importantly, the contributors review the strategic policies proposed to reduce U.S. vulnerability resulting from our dependence on foreign oil.

In Part I, Alice Rivlin summarizes the findings of "The World Oil Market in the 1980s: Implications for the United States," an extensive background paper prepared by the Congressional Budget Office (CBO) at the request of the United States Senate Committee on Energy and Natural Resources. It begins with the observation that the United States imported 42 percent of its total oil supply in 1979 and then goes on to forecast that oil imports could increase to 52 percent by 1985 and 57 percent by 1990 if the present trend continues. The CBO predicts that U.S. oil production will inexorably decline during the 1980s as reserves are depleted, thus making it more difficult to reduce the volume of oil imports. Moreover, the effort to substitute more expensive alternative fuels or energy sources for imported oil will impose an additional burden on the U.S. economy.

Consequently, policymakers are faced with a dilemma. The dependence on oil imports poses a distinct set of risks for the U.S. economy, defined as losses or costs that are not included in the price of oil. If imports are not reduced, the domestic economy will continue to be vulnerable to interruptions in the flow of foreign oil as well as to rising prices and other risks. On the other hand, the costs of substituting domestic energy sources for foreign oil will be very high.

Other alternatives are worth considering according to the contributors. For example, rather than reducing oil imports, policy-

makers might choose to accept the inherent risks imports pose and prepare for their consequences through such means as the Strategic Petroleum Reserve or macroeconomic policies to bolster the dollar.
Another option would be to diversify oil sources so as to minimize
the risk of a break in supply, or to link purchases of foreign oil to
U.S. exports. The chapter presents various alternatives to protect
the United States from the effects of rising oil imports under three
kinds of policies: policies to reduce oil imports, policies to offset the
economic losses posed by imports when they occur, and policies to
reduce the risks inherent in any level of oil imports.

In Part II, John J. Schranz, Jr., and the staff of Resources for
the Future argue that understanding the methodological problems involved in making estimates will help clarify some of the uncertainty
concerning oil and gas resources. Most discussions concerning oil
and gas resources in the past decade have been exercises in confusion
and uncertainty. This quandry is in part the result of a variety of
methodological problems beginning with the lack of clarity in terminology. Another problem involves confusion of estimates with actual
measurements. The proliferation of estimates from various sources
for divergent uses causes another set of problems. Finally, the inap-
propriate use of estimates to draw oversimplified conclusions leads
to even further confusion.

The terminology "oil and gas resources" can be applied to many
different concepts, consequently, the first step is to define the terms.
It can apply to the total resource base, that is, all unproduced natural
oil and gas hydrocarbons that may exist, or it can apply to more limited concepts such as discovered sources. Resources also refer to
reserves, simply proven reserves, or it may include unproven reserves. This term is further complicated by the inclusion or exclusion of subeconomic resources that will vary according to price and
technical changes. The second step requires a clear understanding
by the analyst that there is no actual measurement of any oil or gas
resource, whether discovered or not, proven or not. The only measurement is of production. The produced oil and gas is used to estimate the size of the resource. This estimate of proven reserves
provides information for an estimate on unproven reserves. Finally,
estimates are also produced from various sources for an array of
purposes such as acquisition of loans, public relations purposes, or
to satisfy government regulations. To further complicate matters,
estimates are made by geologists, engineers, and economists using
different techniques and each technique has its own uses and limitations. By knowing the source and purpose of the estimate, the energy
analyst can reduce some of the uncertainty.

It is important to understand that analysts use historical information to provide estimates of future production of oil and gas in a

changing environment and there will always be some uncertainty. The reader will have a better understanding of the problems involved in estimating oil and gas resources after reading this chapter and such an understanding will help prevent the misuse of information.

"The Geopolitics of Oil" is a summary of the testimony of former Central Intelligence Director (CIA) Stanfield Turner before the United States Senate Energy and Natural Resources Committee in April 1980. It provides an excellent overview of the economic, political, and military implications of the world energy situation in the 1980s. The CIA has concluded that world oil production is probably at or near its peak and will decline throughout the 1980s. In this decade, nonoil energy sources are not likely to offset the projected slippage in oil production. Natural gas supplies will increase only slightly, if at all, during the 1980s. If energy supplies remain constant in the United States during this decade, achieving a 3 percent economic growth rate would require that by 1990 the rate of fuel consumption per unit of GNP be reduced by one-third.

The peaking of world oil output and the competition for available supplies by the West, the less-developed countries (LDCs) and soon the Soviet Eastern bloc all add up to less Middle East oil for the United States. Politically, the cardinal issue is how vicious the struggle for scarce energy supplies will become. This competition will create a severe test of the cohesiveness of both the Western and Eastern alliances. Developing forms of cooperation among oil-consuming countries to regulate this competition and to prevent it from becoming mutually destructive will constitute a critical challenge to policymakers and nations.

Part III focuses on two forecasts of Middle East oil supplies to the United States and then examines the energy security issues provoked by dependence on foreign oil supplies. Both forecasts contain political caveats that reduce the confidence of the predictions and remind us that the interjection of politics into the Persian Gulf oil markets adds risk to any long-range plans to use these oil resources.

Fereidun Fesharaki observes that the formation of OPEC was more than the establishment of a supernational organization. It reflects the change in the oil market power balance and the assertion of the sovereign rights of oil-producing nations. He argues that this change is not a temporary phenomena; it is here to stay. While no unified production policy exists within OPEC, the decisions and signals from different nations are interpreted by other exporters so that production is curtailed once a glut develops. The lag time for market adjustments ranges from 6 to 12 months.

Fesharaki also maintains that Western analysts have traditionally forecasted OPEC production by assuming that OPEC would produce to fill the gap left by other producers to satisfy the world's energy demand. This method of forecasting OPEC production is a poor one and is misleading because it fails to consider the economic, social, and political factors that also affect decisions about production levels. Moreover, it fails to consider the possibility that OPEC policy may determine the development of other energy resources and not vice versa. Fesharaki believes there is only one way to project OPEC oil supplies, that is, by counting each member nation's production level, then adding them all up. There are too many factors to consider to estimate production any other way.

In the future, OPEC may perform a number of functions beyond setting production levels and prices. OPEC could become a forum for representing member interests in a range of areas from relations with industrial countries to economic assistance policies toward the developing world. The organization could also serve as a trading bloc to obtain concessions for itself and other LDCs.

Tom E. Burns observes that pessimistic oil forecasts are now as widespread as optimistic ones were a few years ago. The Standard Oil Company of California oil forecast, which he presents, is probably more optimistic than the CIA analysis but certainly is more pessimistic than others. It should be noted that Standard's forecast for the year 2000 does not take into account short-term discontinuities, like the one that followed the revolution in Iran. Nevertheless, in his opinion, the world does seem to be adjusting to the longer-term trend of slower growth in energy use and higher prices.

Burns predicts that by the year 2000, the U.S. energy consumption will increase by one-third, Japan's by 100 percent, and Europe's by an intermediate amount, while the developing areas will have higher energy growth rates than industrialized areas. Because the production of conventional oil will peak in 1990, its share of total energy production will fall from 63 percent to 50 percent in the year 2000. Production of natural gas liquids will continue to increase, but most of the increase in energy production will come from nuclear and coal sources.

In view of this scenario, Burns argues that after 1990 the development of synfuels will compensate for falling conventional oil production in the United States. He forecasts worldwide synfuels production at 4.3 million barrels a day in the year 2000, of which 30 percent will be derived from coal, 30 percent from tar sands, 25 percent from shale, and the rest from biomass. While Exxon's synfuels forecast is twice as high, he observes that achieving even Standard's forecast will require a substantial commitment to synfuels production.

In the United States, Burns believes that conventional oil production may have peaked in 1978 along with gasoline consumption. Synthetic fuels production will reach 1.6 million barrels a day by the year 2000 with two-thirds of it derived from oil shale, but this will barely offset the decline of conventional production. The U.S. production decline means that the United States will remain dependent on foreign oil.

Thomas L. Neff takes the position that the revolution in Iran served to crystallize major changes in the world oil market. Before 1973, 90 percent of OPEC's oil was produced, exported, and traded by major oil companies, which sold the excess to each other or to third parties. As a result, producers in 1973-74 attempted to embargo specific nations, but the oil companies shifted supplies around and compensated for the OPEC shortages. During the crisis in Iran in 1979, consumers paid premium prices for oil and both OPEC and non-OPEC producers began to impose conditions on oil sales, including restrictions on resales and the destinations of shipments. More recently, producers have also begun to make direct spot sales at higher prices, sometimes by reneging on previous contract agreements. As a direct result of these changes, the world oil market is far less flexible and resistant to disruption than was the case previously. The increasingly heavy character of the average crude oil on the world market, which is unsuitable for many refineries, even further constricts market flexibility.

Neff does find some reason to be optimistic about oil supplies. First, smaller trading companies, presumably less subject to produce imposed restrictions, have emerged. Second, producers now find it harder to act in concert. On the other hand, any oil producer can now cause a disruption, which will cause a permanent jump in nominal oil prices or a sort of ratchet effect. The United States' best response to the changing international oil market is to increase the flexibility of the U.S. system during oil interruptions. For example, stockpiling and developing surge production capacity can reduce panic and minimize the ratchet price effect. Saudi Arabia has called on us to conserve oil. Therefore, we might ask them to reserve production capacity in return. The United States should also retrofit refineries to use a broader range of crudes and be prepared to trade Alaskan crude for lighter crudes that domestic refineries can process.

Three critical energy sources for the United States are examined in Part IV: Mexico, Venezuela, and Indonesia. As the United States attempts to diversify its sources of oil and gas supplies from around the world to reduce its vulnerability, energy sources outside the Persian Gulf take on a new significance. By diversifying to maximize flexibility, the United States will increase its presence in other parts of the world. As such, the next two decades will undoubtedly bring an increase in trade between industrialized and developing countries as energy resources are exchanged for highly technical equipment, technical assistance, financial aid, and other goods and services.

Gary J. Pagliano observes that Mexico's proven reserves are about 50 billion barrels of oil equivalent, or the fifth largest in the world. The actual size of the reserve may be as high as 70 billion barrels. Like many other oil-exporting countries, Mexico will not maximize short-term oil production because of political and economic constraints. The nation needs oil revenue to counter three interrelated problems: its rapid population growth, its increasing agricultural imports, and its serious unemployment. PEMEX, the national oil company, has become a symbol of national independence and oil is considered part of the nation's patrimony. Mexico has been willing to tie oil deals to trade and other economic arrangements. These development revenues are likely to be diverted to industries that are more labor intensive and Mexican oil production will probably increase to about 2.5 million barrels a day. After failing to reach agreement with the United States over sales of natural gas, the Mexican government decided to dramatically increase domestic gas use. Unfortunately for the western United States, Mexican gas exports will total only 2.8 billion cubic feet per day by 1988.

David Ronfeldt agrees that Mexico will hold production levels down, but argues that it would be against the United States' own best interest to encourage Mexico to increase production. He believes that Mexico will interpret interference in energy policy as an attack on the sovereignty of the nation because PEMEX is a political institution and has embodied the essence of Mexican nationalism, national dignity, economic independence, and state sovereignty. On the other hand, Mexico is unlikely to join OPEC because this would interfere with the nation's freedom of action.

Ronfeldt concludes that U.S. interests in Mexico go far beyond energy. Mexico is a neighbor whose economy, culture, and society are becoming increasingly intertwined with our own. Pressure from the United States to increase Mexican oil production, or the proportion of Mexican oil exports flowing to the United States, would probably be resented and therefore be counterproductive. More importantly, the high revenues associated with high production levels could further destabilize Mexico's economy and society. In view of this, Mexico's leaders are likely to choose a moderate production rate, on balance most advantageous to U.S. interests, that is, 3.5 million to 5.5 million barrels a day. By the mid-1980s Mexico will probably have surge capacity available for emergencies, but because of concern for national sovereignty, Mexico is unlikely to make bilateral arrangements with the United States. One possible strategy for the United States would be to retrofit domestic refineries to handle heavy Mexican crudes.

Arturo Gandara observes that since 1938 Mexico has followed a consistent hard energy path, emphasizing self-sufficiency through conventional resources including natural gas, petroleum, hydropower, geothermal, coal, and, more recently, nuclear power. The government controls energy policy and resources through the Secretariat of Patrimony. PEMEX, the predominant energy institution, is nominally autonomous, but its board of directors, drawn from government and labor, is appointed by Mexico's president. Other Mexican energy institutions are emerging and most favor the development of nuclear power. He predicts energy consumption will grow at 8 percent annually, partly because oil prices are below world levels.

Frank Tugwell predicts that Venezuelan oil production will decline slowly through the 1980s, from about 2.2 million to 1.7 million barrels a day by the late 1990s. Current domestic consumption is 300,000 b/d; and because of low domestic prices, gasoline consumption is growing by 11 percent a year. The nation's history is a prime example of how the appetite for oil revenue can grow in an open political system. Venezuela has been and will continue to be a responsible hawk on oil prices and its leaders view the nation's extensive heavy oil resources of the Oronoco Belt as a competitor with U.S. synthetic fuels. Production probably will not be substantial until the 1990s. The oil is so heavy and contains so many contaminants that production costs may be in the neighborhood of $35 a barrel. Venezuela nationalized its petroleum industry in 1976 and can afford to purchase the technology to develop heavy oil. However, the Venezuelan government probably would not be receptive to letting American companies invest in the development of the resource.

Guy Pauker argues that because of rising domestic demand and depleted reserves, Indonesia may be forced to cease oil exports sometime in the 1990s. Currently, Indonesia has shown no inclination to limit oil production that is now at about 1.56 million barrels a day and will probably remain a willing supplier. Barring major new discoveries, Indonesia will exhaust its estimated reserve of 10 billion barrels by the end of the century. Meanwhile, domestic oil consumption will increase by at least 12 percent annually as more of Indonesia's 140 million people begin to purchase commercial energy. To avoid the destruction of the nation's forests, which are being stripped for firewood, the government has encouraged the use of kerosene by keeping the price low. Future efforts to increase domestic consumption of natural gas, coal, hydropower, and nuclear power will be hampered by the distance between resources and population centers.

Fortunately for U.S. markets, Indonesia's natural gas resources will be available for export. Japan has been much more aggressive than the United States in building facilities to handle this gas. The Indonesians have held liquefied natural gas (LNG) in reserve for sale to markets in the Western United States possibly because they do not want to become too dependent on Japan as a customer. Other Asian nations are unlikely to make up the gap left by Indonesia's reduction in oil and gas exports. For example, Malaysia's reserves are much smaller, estimated at about 3 billion barrels, while China's reserves, about 20 billion barrels, are larger than Indonesia's. However, all of the oil produced by China will be needed domestically.

Part V presents three assessments of North American energy supplies available to the United States. Because of domestic experiences in 1973 and 1979 when fuel stocks had to be shifted from region to region in the United States to cope with the shortages, these supply sources have also taken on new importance for U.S. policymakers. In addition, regional interests both inside and outside of the United States often conflict with the national government's oil and gas policies.

Arlon Tussing assesses the potential for Alaskan oil production and presents a practical schedule for its development. He predicts that oil production will probably crest at about 1.9 million barrels a day in the mid-1980s and fall to 800,000 barrels a day by the year 2000. The escalation in world oil prices following the Iranian revolution, the federal government's special treatment of Alaskan oil under the price control program, and the legal prohibition on exports to the Far East all combine to create an unexpectedly powerful incentive for West Coast refiners to use Alaskan oil. For example, approximately 40 percent of California's total petroleum supply came from Alaska in 1980. While a break in the Trans-Alaskan Pipeline could be repaired in several days, the destruction of a pumping station by natural causes or sabotage could take from a few months to a year to repair. This heavy reliance on Alaskan oil makes the western United States vulnerable to disruptions, particularly if shortages exist elsewhere in the world at the time of the disruption.

R. D. Hall notes that Canada's hydrocarbon resources are adequate to achieve self-sufficiency, although some adjustments in consumption would be necessary. National policymaking in Canada is complicated because the provinces maintain substantial control over the resources. Current production is 1.85 million barrels a day of liquid hydrocarbons, but reserves of conventional oil have fallen to 5.9 billion barrels. It will be necessary to develop offshore resources and the bitumen resources of Alberta to compensate for these projected shortages. On the other hand, the natural gas outlook is much brighter and supplies will be adequate to allow for exports of 14.2 trillion cubic feet per year. The midwest and western regions of the United States will become strong markets for this clean-burning fuel during the 1980s.

Dennis Eoff describes oil resources, reserves, and production in California, which is currently the fourth largest producer of oil in the United States. California currently produces approximately 950,000 barrels a day, which accounts for 40 percent of the state's total use with the remainder coming primarily from Alaska and Indonesia. Future oil production within the state will be expensive and would have to come from two sources: thermally enhanced oil and offshore outer continental shelf production.

There are approximately 4.2 billion barrels of proven reserves in California defined as heavy oil. Approximately 2 billion barrels of the reserves can be recovered using thermal enhancement techniques. California has been well explored offshore so large new discoveries are not expected. The offshore oil finds are expected to produce approximately 50,000 barrels a day for the next 20 years. An increase in offshore production would reduce U.S. vulnerability to foreign oil shortages. However, the relatively small size of the discoveries and the environmental impacts of oil production in this area make it a less significant source to meet future U.S. oil demands.

Part VI examines the strategic policies developed to reduce the U.S. dependence dilemma. During the past two years, we have witnessed a dramatic shift in the policy debate regarding the proper response to a fuel shortage. Coupon rationing was the focus of the early debates. The problems involved in a system that would create more coupons than the current U.S. currency supply led many to conclude that such an oil crisis management mechanism was unworkable. Rejection of coupon rationing by most policy analysts shifted the focus of the debate to the economic free market price as the preferred oil crisis management mechanism. Most analysts agree that the market can manage a minor oil shortage, defined as a nationwide 5 percent shortfall. But major problems arise during a severe shortage, 20 percent or greater nationwide. Consequently, some type of revenue rebate mechanism is needed to recycle dollars back into the economy. The nature of the proper rebate mechanism and when it should be implemented has not been established. Indeed, the contributors in this part argue that the United States is not even prepared to manage a major oil shortage.

David A. Deese observes that future disruptions of foreign oil supplies are highly likely and the federal response to date has been inadequate. He recommends that each state take some measures on its own while pressuring the federal government to develop a pragmatic plan. Conservation in the United States will help somewhat because it decreases wealth transfers in the event of a disruption. It also helps to create excess capacity in the world, cuts production from unstable areas, and aids the international negotiating position of the United States. However, he concludes, conservation alone is not sufficient.

According to Deese, energy security means physical access to oil not only for the United States but also for our allies at a price that does not create intolerable disruptions. As such, energy security for the United States and its allies is the cornerstone of a sound foreign policy. Given this definition, how should we handle disruptions? Deese argues first that the United States should rely on the price system rather than on an administrative allocation system. Second, the United States should maintain the existing windfall profits tax in some form and third, accelerate the decontrol of oil. Fourth, a federal tax on gasoline with rebates to be enacted only during disruptions should be imposed. Fifth, these measures should be put in place before a disruption occurs. Sixth, the government should ensure that the federal tax does not cause an abrupt jump in the consumer price index. Seventh, different measures should be implemented at different stages and levels of disruption. Finally, the United States must develop a policy to control price as well as access to oil supplies.

Deese offers a number of concurrent specific actions the states and the federal government could take before a crisis arises. The western states could take as much oil as possible from Indonesia and send Alaskan oil to the eastern states. States also should diversify their oil supply before a crisis or have plans in place to do so during a crisis. The United States should be prepared to divert some Alaskan oil to Japan, although this will create a tremendous negative public reaction from the western and midwestern states. Stockpiling is important but raises several questions. For example, when are the stocks to be used? How can we be sure private stockpilers will release them when needed? At what price will the stocks be sold? What blend of crudes or products should be stockpiled? Although emergency conservation plans are not going to be very helpful nationwide, individual states may benefit if the plan is set up well in advance of a crisis. Deese also argues that an emergency tax with a rebate involves fewer difficulties than coupon rationing.

William B. Taylor, Jr., describes four ways of dealing with unexpected disruptions of oil supply. He suggests that we could do what we did last, namely, control oil prices at all levels, allocate gasoline among suppliers, and allow lines to form. The government could institute coupon rationing with a "white market" for coupons, even though accompanying problems include reimposing price controls and issuing what amounts to a second currency. An emergency gasoline tax with a rebate could be imposed. Although this is similar to rationing, it is easier to administer. All oil prices could be decontrolled and windfall profits could be collected.

Carlyle Hystad explains that the federal Strategic Petroleum Reserve (SPR) was supposed to store 150 million barrels of oil in 1978 and 500 million barrels by the end of 1982. The drawdown capacity was planned at 3 to 3.5 million barrels a day, but the Carter administration imposed more ambitious and less realistic goals of 250 million and 500 million barrels by the end of 1978 and 1980 respectively, and 1 billion barrels as soon as possible thereafter. In 1980

SPR capacity was 250 million barrels, with 91 million barrels actually in storage and drawdown capacity of 1 to 1.5 million barrels a day.

Hystad predicts that expansion of existing storage sites will expand capacity to 500 million barrels, but completion could be delayed until 1987-89. The project has been delayed in the past by the formation of the U.S. Department of Energy, which stalled procurements, and by the overemphasis on holding costs down. Any savings were negated by the rising cost of oil. According to Hystad, the most critical problem is a political one, that is, there is still no agreement on when and how the strategic reserve should be used. The SPR should probably be reserved for true national security emergencies and we should look to the private sector to store oil for smaller shortfalls.

Under the Energy Policy and Conservation Act (EPCA), the Department of Energy has the authority to require refiners to store up to 3 percent of their annual throughput. The Department of Energy has rejected the idea, partly because private storage would be more expensive than government storage. Perhaps a more important reason for rejection is the problem of verification that the oil is actually being stored or being drawn down. One workable option is a private storage company, paid for by a tax on the oil industry, that could avoid government procurement delays. With private oil storage, the corporation's board of directors would decide independently when to draw stocks down. Still another option would be to require all oil companies to draw stocks down under federal order with a penalty for noncompliance. Compliance could be measured by monitoring oil input and output. Verification of stock building would be unnecessary.

Susan L. Missner observes that European nations have had more success with private stockpiling than the United States has had with the Strategic Petroleum Reserve. She argues that if the federal government applied the Energy Policy Conservation Act's 3 percent rule to both oil imports and domestic production, it would result in storage of about 180 million barrels. Between the spring of 1979 and the spring of 1980 domestic oil companies increased stocks from 100 days' worth of imports to 153 days. This 800 million barrel inventory is worth $32 billion at 1980 prices, but in a crisis it could be worth $200 billion. The increased costs are presumably passed on to energy consumers, to some extent internalizing the cost of dependence on uncertain oil. Private stockpiling will be less than optimal, however, because no individual firm can capture the benefits of a national oil stockpile and the applicable discount rate is arguably lower for society than for individual firms. Missner believes that this justifies subsidizing private stockpiling.

The advantage of the federal Energy Policy Conservation Act approach is that firms would have an incentive to minimize the costs of required storage. Since all firms would face the requirement, none would be at a competitive disadvantage. Requiring stockpiling of a representative slate of crude or product would prevent companies from storing only the cheapest grades. The Department of Energy could allow firms to store their oil in unused Strategic Petroleum Reserve facilities. Costs would then be lowered and equipment for filling and emptying the reserve could be tested. This method would also separate emergency stocks from regular operating stocks. Otherwise, other time, companies might begin to use emergency stocks in normal operations.

According to Missner, probably the most equitable way to induce stockpiling is through the creation of a central storage corporation based on the German model. Under this system all refineries are obliged to join the corporation, which borrows money, purchases oil as collateral, and arranges for storage facilities. The refiners then pay storage fees, but do not have to tie up their capital in storage facilities.

While it is clear that there is no single remedy to the energy crisis that confronts the United States, many of the contributors in this volume go beyond mere problem definition and recommend possible solutions. After reading these divergent viewpoints, we hope that the reader has redefined the terms of the energy debate, has developed some insights into the complexity of the issues, and developed an appreciation for the need for the United States and other nations to cooperate as well as compete for scarce energy resources. We view this text as an introduction to a set of challenges to be met jointly by the public and private sector over the next 20 years.

The realization that there are no measurements in oil and gas resource appraisals is important to impress upon everyone. Even in discovered reservoirs we do not measure the oil in place—it is estimated. Reserves are an estimated value derived from a prior estimate of the oil in place, taking into account economics and technology. If the oil-in-place estimate changes, so will that of the reserves. Reserves and resources are equal to the estimated oil in place multiplied by an assumed recovery factor, substantially less than 100 percent for oil, less the amount of cumulative production. Nothing could be simpler yet so uncertain. There are no hidden formulas for predicting the end of the finite supply of oil and gas in the United States or the world. The definitive study of future oil and gas supply and how it may be altered by economic and technological parameters that have not yet emerged still remains to be done. It is of little comfort that the final, reliable appraisal of the oil and gas resources of the United States will prove to be historic rather than predictive.

I

The United States and World Energy Sources

INTRODUCTION

Since the disruption of oil flows in 1973, public policymakers in the United States and throughout the world have struggled to develop an energy policy to meet the challenges posed by the realities of the world oil scene. A common threat through all of the legislation enacted by the United States Congress has been the goal of reducing oil imports by substituting additional domestic sources of energy. Even with the substantial efforts of the 1970s and those of the 1980s, policymakers in the United States and other oil-dependent nations are still faced with dependence on imports and the numerous economic and political risks and problems associated with dependence. Many experts agree that these risks include:

- Vulnerability to interruptions of vital supplies
- Long-term price increases
- Constraints on foreign relations
- Negative impact on the strength of the dollar

The severe shock dealt the United States and many nations of the world during the 1979 Iranian revolution was a stark reminder of the risks associated with dependence. It is this uncertain and potentially dangerous environment that all public policymakers face in the 1980s.

In Chapter 1, Alice Rivlin summarizes an extensive paper prepared by the Congressional Budget Office (CBO) at the request of the United States Senate Committee on Energy and Natural Resources. Although it is very difficult, if not impossible, to forecast accurately the energy picture for a decade or more in the future, the chapter clearly delineates the myriad of energy problems facing the world and the dilemmas facing policymakers and citizens as solutions to energy problems are sought.

1

THE WORLD OIL MARKET IN THE 1980s: IMPLICATIONS FOR THE UNITED STATES

Alice Rivlin

INTRODUCTION

In response to sharply increased world prices and the instability of oil imports, the Ninety-fifth and Ninety-sixth Congresses passed several pieces of crucial energy legislation, putting a national energy policy into place. Natural gas and petroleum product fuel prices will now more accurately reflect their market levels. A program to develop synthetic liquid fuels and gas, financed in part by the federal government, has been started, and firms engaged in this development may be given expedited regulatory treatment through an Energy Mobilization Board. Tax credits and grants are being used to subsidize conservation of energy, and low-income families are being offered assistance adjusting to higher energy prices. A program for gasoline rationing is under development for use in the event of emergencies, and the Strategic Petroleum Reserve may resume purchases. Revenues from a "windfall profits" tax may be available for many of these projects.

Despite these actions, the oil import problem remains unsolved, partly because of declining domestic production and a persistent demand for oil. Barring significant price increases or new policy decisions, another decade of oil imports lies ahead. By the end of the 1980s, the United States may be importing more oil than it did at the beginning of the decade.

The U.S. dependence on oil imports poses a distinct set of risks for the U.S. economy. These risks may be identified as economic losses or costs that are not included in the price of oil. They are of uncertain magnitude and related to events that may, but not necessarily will, occur. The extent of the risks is governed by both the level of imports and the state of the world oil market. The flow of imports, for example, can be disrupted for both political and logistical reasons, and disruptions larger than those experienced in 1973 and 1979 are possible. Such disruptions would depress the growth of the gross national product (GNP) and lower the U.S. standard of living. The price of imports may continually rise, dampening the long-term prospects for economic growth and increasing inflation. Imports create dollar outflows unmatched by the demand for U.S. goods, services, and assets. These depress the value of the dollar and may jeopardize the world monetary system. U.S. relations with other countries may be affected by the nation's linkage to foreign oil.

While policies to reduce oil imports may be successful, lower import levels will probably not be realized until the 1990s. Moreover, regardless of their success, import reduction policies do not confront directly the risks that oil imports will pose, even at reduced levels.

Congress will, therefore, have to decide if a "second round" of energy policy is necessary, and if so, what form it should take. Making these judgments will require an appraisal of the oil import problem over the next decade—an appraisal not only of the future level of imports, but also of the risks these imports will present for the United States. This chapter analyzes each of these risks individually, and discusses the array of possible policy responses to them.

PROJECTIONS OF U.S. SUPPLY AND DEMAND FOR OIL

U.S. imports of crude oil and refined products increased dramatically in the last decade, and they are projected to continue to rise in the coming decade although at a more moderate rate. From 3 million barrels per day in 1969, imports grew to 6 million barrels per day in 1973 and 7.8 million in 1979. Assuming that oil prices rise by 2 percent annually in real terms and that current federal policy remains unchanged, the level of oil imports is estimated to reach 10.1 million barrels per day in 1985. By 1990, this level may reach 11.3 million barrels per day. The projected import levels represent the difference between estimates of future demand for oil in the United States and estimates of the supply of domestic oil.

The Demand for Energy: An Overview

The U.S. demand for energy from all sources will continue to grow during the 1980s, reaching the equivalent of 47.8 million barrels of oil per day by 1990, a 22 percent increase over total energy demanded in 1979 (see Table 1.1). This is an average annual increase

TABLE 1.1

U.S. Energy Demand by Primary Energy Source
in 1979, 1985, and 1990
(in millions of barrels per day of oil equivalent)

	1979 (estimated)	1985 (projected)	1990 (projected)
Amount			
Oil	18.7	19.5	19.9
Natural gas	9.8	9.5	9.7
Coal	7.6	10.0	12.2
Nuclear energy	1.5	2.6	3.9
Hydropower	1.5	1.5	1.5
Other*	0.2	0.3	0.6
Total	39.3	43.4	47.8
Percent distribution			
Oil	47.6	44.9	41.6
Natural gas	24.9	21.9	20.3
Coal	19.3	23.0	25.5
Nuclear energy	3.8	6.0	8.2
Hydropower	3.8	3.5	3.1
Other*	0.5	0.7	1.3
Total	100.0	100.0	100.0

*Includes solar and other renewable resources.
Source: Congressional Budget Office, using Data
Resources (DRI) energy model, December 19, 1979.
See the original report for a thorough discussion of the
methodology used in the preparation of study: The World
Oil Market in the 1980s: Implications for the United
States, The Congress of the United States, Congressional
Budget Office, Government Printing Office, Washington,
D.C., May 1980, pp. 5-6.

of 1.8 percent a year, as compared with the higher rate of 3.7 percent a year for the postwar period.

The domestic demand for oil will also continue to grow gradually
during the next decade, rising from 18.5 million barrels per day in
1979 to 19.5 million in 1985 and 19.9 million in 1990. But the share

of oil as a source of primary energy is projected to decline substantially, from 47.6 percent of the total in 1979 to 41.6 percent in 1990.

This slowdown in the rate of growth of domestic demand for energy and the decline in the share of oil as a primary energy source can be attributed to several factors. First, demographic trends suggest slower increases in the labor force, in persons of driving age, and in new households. Second, the 1980s are likely to see a slower rate of growth in the GNP, and this will dampen domestic demand for energy. Third, structural shifts now underway in the composition of the economy, stressing less energy-intensive processes and new technologies, suggest that the demand for energy will grow at a slower rate than output and that this gap will increase over time. Fourth, price increases, especially for oil, will provide an incentive for both households and businesses to conserve oil. Finally, recent and future energy price increases may make unconventional sources of energy economically viable, thereby encouraging their development and use.

Sectoral Demand for Energy and Oil

While the demand for oil is projected to increase gradually over the 1979-90 period, the share of oil as a source of energy is projected to decrease. At the same time, demand for coal is projected to grow from 7.6 million barrels per day of oil equivalent in 1979 to 12.2 million in 1990, an annual rate of increase of 4 percent. Demand for nuclear power is projected to increase from 1.5 million barrels per day of oil equivalent in 1979 to 3.9 million in 1990 as existing projects are completed, an annual rate of increase of 9.1 percent (see Table 1.2).

The rates of growth in energy demand, and in demand for oil in particular, are projected to vary in different sectors of the economy. In transportation, conservation efforts will slow the growth of oil consumption. In the residential and commercial sectors, the substitution of electricity and natural gas will reduce the growth in oil consumption between now and 1990. Electric utilities will also reduce their use of oil as nuclear energy and coal become more important sources of primary energy. Consumption of energy in manufacturing will remain relatively constant, as will the sector's relative emphasis on oil as a primary energy source. Over the next decade, other fuels will be increasingly substituted for oil—first, coal and nuclear energy. In the longer run, solar energy and some synthetic fuels will become viable economic alternatives. It should be noted, however, that opposition to nuclear power and coal burning will limit their use. In the absence of this opposition, these two sources could replace an additional 1 million barrels per day of oil in 1990.

Residential and commercial primary energy consumption is

TABLE 1.2

Total U.S. Energy Demand and Oil Demand by Sector of the Economy
in 1979, 1985, and 1990
(in millions of barrels per day of oil equivalent)

	1979 (estimated)	1985 (projected)	1990 (projected)
Total energy demand			
Residential and commercial	7.5	8.6	9.2
Industrial	9.8	10.6	11.1
Transportation	9.8	9.9	10.3
Electric utility	12.1	14.3	17.4
Total	39.2	43.4	47.8
Oil demand			
Residential and commercial	3.5	4.1	4.2
Industrial	3.7	3.9	4.2
Transportation	9.8	9.9	10.3
Electric utility	1.7	1.6	1.0
Total	18.7	19.5	19.9
Oil as a percent of total energy			
Residential and commercial	46.7	47.7	45.7
Industrial	37.8	36.8	37.8
Transportation	100.0	100.0	100.0
Electric utility	14.0	11.2	5.7
Total	47.7	44.9	41.6

Source: CBO, using DRI energy model, December 19, 1979.

projected to grow from 7.5 million barrels per day of oil equivalent
in 1979 to 9.2 million in 1990, an average rate of 1.9 percent a year.
Oil consumption will rise from 3.5 million barrels per day to 4.2
million barrels per day during this period. The leveling off in oil
consumption will be the result of conservation measures in space
heating and cooling, increased use of electricity, development of
more efficient electrical systems, and a decline in the rate of new
household formation.

The industrial sector will show only moderate growth in energy consumption—from 9.8 million barrels per day of oil equivalent in 1979 to 11.1 million in 1990—and the share of oil in total energy consumption will remain roughly constant at 37.8 percent. The most important factor in industrial energy demand is overall economic activity as reflected in the GNP. Substitution of other fuels has not been extensive in the past for a number of reasons. Curtailments in the supply of natural gas for industry have forced some substitution to oil. Moreover, until recently, energy from synthetic fuels has not been an economic alternative. As the real price of oil continues to rise, however, synthetic fuels may be utilized. Coal is growing in importance to the industrial sector despite obstacles to more rapid growth, such as lack of available land for coal storage, the requirement of clean burning in some industrial processes, and air pollution standards.

The transportation sector currently accounts for more than one-half of U.S. oil consumption and one-fourth of all energy consumption. Oil use in transportation is projected to grow only slightly, from 9.8 million barrels per day in 1979 to 10.3 million in 1990. This slower growth will be mainly a reflection of conservation as the fuel economy standards for autos of the Energy Production and Conservation Act (EPCA) take effect. EPCA efficiency gains may be less than anticipated, however, if consumers continue to shift to the less energy-efficient portion of the new transportation fleet—light trucks and vans. Moreover, gasoline prices will have to rise further before many gasoline-saving innovations become cost-effective. Since the transportation sector is almost exclusively dependent on petroleum, little substitution of other fuels may be expected in the short run. Technological innovation will be required before electric vehicles make significant inroads in the automobile market.

Electric utilities will continue to be the fastest growing energy sector of the economy. Energy consumption by utilities will grow from 12.1 million barrels per day of oil equivalent in 1979 to 17.4 million in 1990, an average rate of 3.4 percent a year. Since fuel is their main cost, electric utilities have always had an incentive to improve energy efficiency, although some rate-setting procedures may work against this. New technologies, combined with conservation measures to utilize waste heat, promise some energy savings in this sector. Oil is expected to continue to decline in importance as an energy source, from 14.0 percent of fuel used in electric utilities in 1979 to 5.7 percent in 1990. On the other hand, coal and nuclear energy will become increasingly important.

Domestic Oil Supply

Domestic crude oil production is projected at 8 million barrels per day in 1985 and 7.4 million in 1990. It will decline despite the strong incentives provided by the decontrol of crude oil prices because of limited geological prospects and a continuing fall in the rate at which exploration actually results in new reserves.

Production from proved reserves will constitute the bulk of production from the lower-48 states until the late 1980s. At present, two-thirds of it comes from 200 relatively large fields, most of which have been producing for 10 years or more and half for 25 years or more. Conventional primary and secondary recovery methods have been used on almost all of these fields, and production levels are declining because of depletion. In the absence of new reserve discoveries, these fields have averaged production declines of 12.7 percent per year under EPCA controls. Decontrol will arrest the decline somewhat by encouraging investments that accelerate production from these known reserves. Primary and secondary production from known reserves outside Alaska is projected to be 3.2 million barrels per day in 1985 and 1.2 million in 1990.

New Discoveries

Two techniques are used herein to estimate discoveries of new or additional reserves. First, additions to existing reserves, such as revisions of estimates of proved reserves, extensions to known fields, and new pool discoveries are individually projected using historical trends during periods when real prices increased. Second, discoveries of new fields are estimated by combining projected finding rates with estimates of future exploratory drilling rates, which, in turn, are directly influenced by prices. Discoveries of new reserves in "frontier areas" are projected by synthesizing estimates from individual firms, trade organizations, and government agencies. Once new reserves are established, a conventional production profile over time is assigned to them.

Domestic production will be influenced more by the decline in production from known fields than by new discoveries until the late 1980s. With the exception of the discovery of Prudhoe Bay, aggregate withdrawals from production have exceeded new discoveries for each of the past 10 years. Although exploratory and drilling activities have increased greatly in recent years and discoveries have increased in number, they have not been large enough to compensate for the depletion of older, larger fields. Unless an exceptional new discovery is made on the North Alaskan Outer Continental Shelf (OCS) or on some OCS areas in the lower-48 states, this trend is expected to continue through the next decade. Production from new reserves is estimated at 2.3 million barrels per day in 1985 and 3.5 million in 1990.

Alaskan North Slope

Alaskan North Slope oil is estimated separately because of its distinct geology and its singular importance in U.S. supply. The producers of Alaskan oil are somewhat unsure of the production profile from the existing Prudhoe Bay fields because of changing economic conditions and because of unanticipated changes in the behavior of the main reservoir. Other North Alaskan fields are not likely to produce more than 300,000 barrels per day until the late 1980s. The Kuparuk reservoir is expected to be the most important of those now known to the public. Total Alaskan production is estimated at 1.6 million barrels per day in 1985 and 1.4 million in 1990.

Heavy Oil and Tertiary Recovery

Heavy oil is defined as oil with an American Petroleum Institute viscosity measurement (API) of 16° or less. This measurement defines a specific gravity of oil that usually renders it less mobile than lighter crude oil. Heavy oil production usually employs a variety of thermal techniques designed to increase mobility so that it will flow when pumped. In some cases, the oil is combusted in the ground to build pressure and heat. Usually, however, oil is burned at the production site in order to generate the steam heat needed to inject into the heavy oil formation. This, in turn, can create significant air quality problems. Most heavy oil reserves are found in California, where stringent environmental standards must be met before production can begin. The decontrol of heavy oil will probably raise the rate of return on production from new fields to the point at which it will be profitable to control emissions and meet state environmental standards. Heavy oil production is estimated at 500,000 barrels per day in 1985 and 600,000 barrels in 1990.

Enhanced, or tertiary, production will also be facilitated by the 1979 price increases of the Organization of Petroleum Exporting Countries (OPEC). With the limited prospects for significant additions to new reserves, extensions of existing reserves through enhanced production will be important in defining the future U.S. reserve base. Chemical injection into formations, previously prohibited by the expense of the chemicals, is now a realistic possibility. Carbon dioxide injection will also become a more feasible technology, given recent price increases. Enhanced production is expected to rise over its current levels by 365,000 barrels per day in 1985 and 690,000 barrels in 1990. Existing enhanced production is counted in production from existing reserves in this analysis.

TABLE 1.3

Projected U.S. Oil Imports, 1985 and 1990
(in millions of barrels per day)

	1985	1990
Projected oil demand*	19.5	19.9
Minus projected domestic oil supply	8.0	7.4
(Proved reserves)	(3.2)	(1.2)
(New discoveries)	(2.3)	(3.5)
(Alaskan North Slope)	(1.6)	(1.4)
(Heavy oil)	(0.5)	(0.6)
(Future tertiary)	(0.4)	(0.7)
Minus projected supply of natural gas liquids	1.4	1.2
Projected oil imports	10.1	11.3

*Crude equivalent of demand for refined products—corrects for increased volume resulting from refining.

Source: Congressional Budget Office.

Projected Oil Imports

In 1985, the U.S. oil demand is projected to be 19.5 million barrels per day, rising to 19.9 million in 1990 (see Table 1.3). The U.S. oil supply in 1985 will total only 8 million barrels per day, and will decrease to 7.4 million in 1990. In addition, production of natural gas liquids will total 1.4 million barrels per day in 1985 and 1.2 million in 1990.[1] If the United States is to meet the projected demand, then oil imports will have to equal the amount over the supply of domestic oil and natural gas liquids, or 10.1 million barrels per day in 1985 and 11.3 million in 1990. These levels are based on a price of $30 per barrel in the fourth quarter of 1979, rising at an annual rate 2 percent faster than the general price level. They also assume that there is no change in existing energy policy.

THE WORLD OIL MARKET AND OIL PRICES IN THE 1980s

Prices in the world oil market are likely to continue rising in the next decade. In the past, such increases have led to increases in domestic energy prices and in the costs of a wide variety of goods and services that require energy inputs. Higher costs dampen consumer demand and may cause a slackening of economic activity, economic growth, and employment, while also increasing the rate of inflation. The ensuing losses of consumer income are multiplied throughout the economy. In the long run, rising energy prices pose a major uncertainty for investors and inhibit the process of capital formation. To assess the likelihood of these prospects, one must analyze the supply and demand trends for oil.

Projections of world oil prices based on supply and demand contain both economic and political uncertainties that make precise estimates unreliable. But supply and demand forecasts imply the likelihood of higher prices, if not their extent. Specifically, these forecasts describe the magnitude of the risk of higher oil prices in this decade, should be world economy achieve moderate growth.

Total world demand for oil (outside of the Soviet bloc and the People's Republic of China) is projected to be 58.9 million barrels per day in 1985 and 66.4 million in 1990. The oil available to meet this demand is estimated at 54.5 million parrels per day in 1985 and 55.8 million in 1990. This equals the total supply of oil from the industrialized nations, including the United States; from OPEC; from developing countries outside OPEC; and from the net exports of the Soviet Union, Eastern Europe, and the People's Republic of China. These projections are based on a fourth-quarter 1979 price of $30 per barrel in the United States, rising at 2 percent above the inflation rate annually. Should they prove correct, the world market will be forced to absorb an excess demand of 4.4 million barrels per day in 1985 and 10.4 million in 1990. Thus, a tight market and rising oil prices, with their potential dampening effect on economic growth, seem likely in the next decade.

World Oil Demand

World oil demand, excluding the Soviet bloc and China, is projected to increase at an average annual rate of 2.2 percent over the 1978-90 period. The largest absolute increases and the fastest rates of growth will be seen in OPEC nations and in the other LDCs. These estimates are presented in Table 1.4.

Demand in the Industrialized Countries

Demand for oil within the industrialized countries of Western Europe, North America, and Japan is projected to rise from its 1978 level of 40.1 million barrels per day to 43.9 million barrels per day in 1985 and 46.8 million in 1990. This reflects an average annual

TABLE 1.4

World Crude Oil Demand, 1978, 1985, and 1990
(in millions of barrels per day)

Area	1978	1985 (projected)	1990 (projected)	Average Annual Rate of Change (percent) 1978-85	1985-90
Industrialized countries					
United States	18.3	19.5	19.9	0.9	0.4
Western Europe	14.6	15.9	17.3	1.2	2.0
Japan	5.4	6.5	7.4	2.7	2.6
Canada	1.8	2.0	2.2	1.5	1.5
OPEC	2.5	4.3	6.3	8.0	8.0
Non-OPEC LDCs	8.5	10.7	13.3	3.3	4.4
Total	51.1	58.9	66.4	2.1	2.4

Source: Congressional Budget Office.

rate of increase of 1.3 percent between 1978 and 1990. An assumption of approximately 3 percent rate of economic growth for the group as a whole was made for this time period, and the demand for all energy was projected to increase by an average of about 2.4 percent annually. Thus, the projections assume a shift away from oil toward nuclear power, coal, and increased conservation.

Higher energy prices are promoting a shift in energy demand in both Japan and Western Europe, similar to that in the United States. In Western Europe, oil demand is projected to increase at an annual rate of 1.4 percent over the period 1978-90. This is lower than the general rate of energy demand growth largely because of the development of alternative energy sources. Between now and 1985, many of the nuclear projects undertaken since 1973 in Western Europe will be completed, taking pressure off oil demand. In addition, new imports of natural gas from a Libyan pipeline under the Mediterranean, and perhaps from the Soviet Union, will further reduce the demand for oil. On the other hand, it is unlikely that new conservation efforts will affect demand growth significantly. Since the OPEC price increases of 1973-74, Western Europe has, by and large, already undergone a transformation to greater energy efficiency in transportation, space heating, and use of waste heat in industry and utilities.

Japanese oil demand is projected to grow more rapidly than European or U.S. demand, because of Japan's anticipated higher rates of real economic growth (4 percent annually, contrasted to the 3 percent assumption made for the United States and Western Europe). This is also true because of Japan's long-term policy of tying oil imports to reciprocal commodity and capital exports, which serves to increase both economic activity and the availability of oil. It will also serve to increase Japanese imports of liquefied natural gas. But Japan's export-oriented economy may not grow as rapidly as was anticipated before the 1979 price increases because of slower economic growth worldwide. Evidence of such a slippage is already appearing in Japan's foreign trade position. To the extent that Japan is forced to import a worldwide recession, its demand for oil will increase less rapidly. Demand levels of 6.5 million barrels per day in 1985 and 7.4 million in 1990 are projected for Japan, reflecting an average annual increase of 2.7 percent over the 1978 level of 5.4 million barrels per day.

Demand in OPEC Countries

OPEC countries experienced an increase in oil demand of 12.8 percent annually over the 1967-78 period as they used oil as a fuel for industrial development and as personal incomes in the OPEC countries rose. While strong future demand increases are projected for OPEC, there is reason to believe that rates will not reach those of the past decade. First, many gas utilization projects have been brought on line that will allow economic use of natural gas that until now has been burned off. Second, the commitment to develop petrochemicals and refining capacity appears to be easing in favor of purchasing participation in existing Western projects instead. Demand within OPEC seems likely to grow at a rate slightly less than that of general economic growth. In addition, the overthrow of the Pahlavi regime in Iran has led to an abandonment of the shah's industrial plans, which were slated to increase Iranian oil demand by 1.5 million barrels per day by 1990. An increase of half as much is now more likely. These factors suggest a long-term growth rate of 8 percent for OPEC oil consumption, raising it from 2.5 million barrels per day in 1978 to 4.3 million in 1985 and 6.3 million in 1990.

Demand in Non-OPEC Less-Developed Countries

Until the 1979 oil price increases, less-developed countries outside of OPEC had been projected to increase their oil demand at an average annual rate of 6 percent or more. The current price of oil has reduced this prospect considerably because of the scarcity of foreign exchange with which to pay for oil imports.

LDCs demonstrated high rates of oil demand increase in the recent past—about the same as general economic growth—primarily because industrial development brought with it new uses of energy. In addition, the migration of workers into metropolitan areas from rural areas meant that their consumption of energy was transferred from indigenous sources used in the country (such as wood, peat, or other natural fuels) to fuels provided by the market. Thus, oil demand probably increased at a higher rate than the total energy demand. This "migration" component of energy demand growth will probably taper off in the 1980s.

In one subgroup of LDCs, rising incomes have created an increased consumer demand for oil. In this subgroup, often referred to as "newly industrialized countries," oil demand is projected to grow at more than 5 percent per year, slightly less than the projected 6 percent rate of economic growth estimated for this group. The largest increases will be seen in Mexico and Taiwan. Australia is included in the group because of its location rather than its level of development. Lesser increases are projected for India and South Korea. Demand may not grow in Brazil, the other member of the group, where 60 percent of foreign exchange earnings are now being returned as debt service to public and private international lenders. Yet all of these countries have export-led economies that are capable of generating foreign exchange.

For LDCs outside this subgroup, demand is projected to increase in proportion to economic growth at an annual rate of approximately 2.5 percent. Many of those outside the industrializing group may be unable to meet oil payments should demand or prices increase considerably. Moreover, slower economic growth in the industrialized community will mean a smaller export market for LDCs, diminishing their prospects for economic growth. Total LDC demand is projected to grow from the 1978 level of 8.5 million barrels per day to 10.7 million in 1985 and 13.3 million in 1990, reflecting a long-term rate of increase of 3.8 percent.

World Oil Supply

The available world supply of crude oil is estimated at 54.5 million barrels per day in 1985 and 55.8 million in 1990. More than one-half will come from the OPEC countries and the remainder from the United States and Canada, the North Sea, and non-OPEC LDCs (see Table 1.5).

TABLE 1.5

World Crude Oil Supply, 1978, 1985, and 1990
(in millions of barrels per day)

	1978	1985 (projected)	1990 (projected)
Industrialized countries			
United States	10.3	9.4	8.5
North Sea (Great Britain and Norway)	1.5	3.7	3.7
Canada	1.6	1.5	1.9
Other	0.4	0.5	0.5
OPEC countries			
Saudi Arabia	8.3	8.5	8.5
Other OPEC	22.0	21.6	22.8
Non-OPEC LDCs			
Mexico	1.3	4.5	5.5
Other Latin American	1.2	1.4	1.2
Africa	1.0	1.8	2.3
Asia	0.9	1.8	1.9
Non-OPEC Middle East	0.6	0.8	0.8
Soviet Bloc and China, net exports*	2.0	-1.0	-1.8
Total	51.1	54.5	55.8

*Includes natural gas.
Source: Congressional Budget Office. For a discussion of the methodology used in developing the data see original report, pp. 18–20.

Supply in the Industrialized Countries

The future supply of U.S. crude oil and natural gas liquids was estimated previously at 9.4 million barrels per day in 1985 and 8.6 million in 1990. Production of industrialized countries other than the United States is estimated at 5.7 million barrels per day in 1985 and 6.1 million in 1990, with the bulk of this from the North Sea fields and Canada (see Table 1.5).

The North Sea reserves are estimated at between 20 billion and 30 billion barrels, depending on the presumed efficiency with which the oil is recovered (see Table 1.6). Expansion of production facilities in the North Sea was temporarily slowed in the late 1970s because of private sector concern with the leasing terms desired by the British government. The North Sea fields are capable of producing 4.5 million barrels per day by 1985 (3.3 million barrels per day from the British tracts and 1.2 million barrels from the Norwegian). The British government, however, will probably put a ceiling on production designed to meet its domestic needs and allow only a minimum for export. A production level of 2.5 million barrels per day has been discussed. This would allow total North Sea output of 3.7 million barrels per day in both 1985 and 1990.

Canada's reserves are set at 6 billion barrels, but are growing because of new discoveries in Alberta. Moreover, the possibility of large new discoveries is high both in the Arctic Islands and in the Mackenzie Delta (near Prudhoe Bay). The Grand Banks area off Newfoundland is thought to have significant potential as well. Canada is also beginning to develop its heavy oil resources. If a recovery technology proves economic, which is now thought to be the case, significant quantities of oil will be available. Canadian heavy oil reserves could total 950 billion barrels. Under favorable conditions, 140 billion barrels out of the 950 billion can ultimately be recovered. Total Canadian production is seen as 1.5 million barrels per day in 1985 and 1.9 million in 1990.

Outside of the United States, the North Sea, and Canada, reserves and production of other industrialized nations are minimal. Production of 0.5 million barrels per day in 1985 and 1990 is seen for the rest of Western Europe, coming from Austria, Denmark, France, Greece, Italy, the Netherlands, Spain, and West Germany. Ireland has the geological potential for large, new fields, but these have not yet been explored.

Supply in OPEC Countries

The OPEC countries will produce more than half of the world's oil for the balance of this century. OPEC production is projected to be 30.1 million barrels per day in 1985 and 31.3 million in 1990 (see Table 1.7).

The distribution of output among the members of OPEC should remain roughly constant over this period. Saudi Arabia, which increased its output after the Iranian revolution, should return to its desired level of 8.5 million barrels per day throughout the 1980s. Although capacity increases could bring output to 12 million in 1990— presuming a continuation of the existing relationship between the

TABLE 1.6

Proved Oil Reserves, January 1, 1980

	Proved Reserves (in millions of barrels)	Reserve Life (in years)[a]
Industrialized countries		
Canada	6,800	12.58
Norway	5,750	40.39
United Kingdom	15,400	26.87
United States	26,500	8.39
Other	2,401	26.96
Subtotal	56,851	12.57
OPEC countries		
Algeria	8,440	18.65
Ecuador	1,100	13.70
Gabon	500	7.13
Indonesia	9,600	16.44
Iran	58,000	54.79
Iraq	31,000	25.20
Kuwait[b]	68,530	75.40
Libya	23,500	31.41
Nigeria	17,400	20.11
Qatar	3,760	21.46
Saudi Arabia[b]	166,480	47.86
United Arab Emirates	29,411	44.15
Venezuela	17,870	21.01
Subtotal	435,591	39.00
Non-OPEC LDCs		
Angola	1,200	22.99
Argentina	2,400	13.99
Australia	2,130	13.26
Brazil	1,220	20.26
Brunei	1,800	19.34
Egypt	3,100	16.99
India	2,600	29.68
Malaysia	2,800	28.41
Mexico	31,250	57.46
Oman	2,400	22.67
Tunisia	2,250	58.71
Other	4,562	12.11
Subtotal	57,712	41.51

TABLE 1.6 (cont.)

	Proved Reserves (in millions of barrels)	Reserve Life (in years)[a]
Soviet Bloc and China		
Soviet Union	67,000	15.73
China	20,000	26.09
Other	3,000	22.21
Subtotal	90,000	17.44
Total	641,624[c]	28.09

[a]Number of years that existing production levels can be maintained, given reserve levels.

[b]Includes shares of "Neutral Zone," an area whose production is divided between Saudi Arabia and Kuwait.

[c]Details may not add to totals because of rounding.

Source: Oil and Gas Journal, December 31, 1980.

United States and Saudi Arabia—growing conservationist sentiment within Saudi Arabia could limit actual production increases.

Some members of OPEC will have to contend with depletion problems. Algeria, Indonesia, Nigeria, and Venezuela all face diminishing production from known reserves, but, with the possible exception of Nigeria, each has a good chance of discovering new resources. Venezuela, with a promising continental shelf and its heavy oil belt around the Orinoco River, is notable in this regard; it should be able to expand production both through these resources and through enhanced recovery in old fields. Iran faces problems of pressure maintenance in older fields, although a gas injection program could maintain its production level. Good prospects also exist for discoveries in the Persian Gulf.

Iraq could be the fastest growing producer in OPEC in the 1980s, increasing output to 5 million barrels per day by 1990. Libya also has the resources to increase output significantly, but may not need the resulting revenue. Algeria, Kuwait, and the small Persian Gulf states should be able to maintain constant output without difficulty, but may also elect to conserve reserves by further limiting production.

Ironically, higher oil prices actually may discourage higher OPEC production levels. The recent doubling of oil prices has brought the revenues of most producing nations to levels that cannot be effi-

TABLE 1.7

OPEC Crude Oil Supply, 1978, 1985, and 1990
(in millions of barrels per day)

Country	1978	1985 (projected)	1990 (projected)
Algeria	1.2	1.0	1.0
Ecuador	0.2	0.3	0.3
Gabon	0.2	0.3	0.2
Indonesia	1.6	1.7	1.7
Iran	5.1	3.5	4.0
Iraq	2.6	4.0	5.0
Kuwait	2.1	1.6	1.6
Libya	2.0	2.0	2.0
Nigeria	1.9	2.5	2.0
Qatar	0.5	0.5	0.5
Saudi Arabia	8.4	8.5	8.5
United Arab Emirates	1.8	2.0	2.0
Venezuela	2.2	2.2	2.5
Total	29.8	30.1	31.3

Source: Congressional Budget Office.

ciently absorbed by economic development. Rather than produce for
a surplus of foreign exchange, many producers may seize higher
prices as an opportunity to leave their depleting oil resources in the
ground.

High-absorber nations. The "high-absorber" OPEC countries—
so named because they are able to absorb relatively high amounts of
petrorevenues through investment and economic development—are
Algeria, Ecuador, Gabon, Indonesia, Iran, Iraq, Libya, Nigeria,
and Venezuela. These countries as a group should provide 17.5 million barrels per day in 1985 and 18.7 million in 1990.

Iran and Iraq are the two largest producers in this group. Iran's
reserves of 58 billion barrels are estimated on the basis of a 15 percent recovery rate applied to an estimated 375 billion barrels of oil
in place. (Only a percentage of actual oil in place is ever recovered. [2]
Reserve estimates are usually made by applying an estimated recovery
rate to the estimated amount of oil in the ground.) If pressure maintenance and other enhanced recovery operations are successful, these

reserves could be ultimately extended up to 100 billion barrels. Pressure to avoid mounting debts and to fulfill expectations of improved standards of living may induce Iran to produce at close to full capacity over the next decade. This could translate to 4 million barrels per day in 1985 and 1990. This level, however, presumes that the Iranian government reestablishes control over its oil fields and has access to spare parts, now denied by the U.S. embargo. While Iranian reserves suggest a higher production figure, the current government may not be willing to invest in the pressure maintenance and exploration necessary to increase capacity to levels reached under the Pahlavi regime.

Iraq's situation is similar to that of Iran, but it has a reserve base that is expected to grow rapidly. The present figure of 31 billion barrels will probably double in the 1980s. Iraq's significant underdevelopment as a producer is explained both by the physical characteristics of its oil (which is heavy, with relatively high concentrations of sulfur and corrosive metals) and by politics. Iraq has had a radical government that antagonized many firms. In 1960 it found only 1 out of every 200 acres leased for oil exploration was actually being worked, and confiscated unworked leases as a response. This has strained its relations with the international oil industry. Iraqi production should reach 4 million barrels per day in 1985 and 5 million in 1990.

The second tier of this high-absorber group is composed of Algeria, Indonesia, Libya, Nigeria, and Venezuela. Algeria and Indonesia both face depletion of known reserves, but should be able to sustain production levels of 1 million and 1.7 million barrels per day, respectively, until 1990. Nigeria faces a similar problem, complicated by the fact that its reserves lie in a large number of small deposits, which makes enhanced recovery operations less profitable. Nigerian production should peak at 2.5 million barrels per day in 1985 and decline to 2 million in 1990, unless unanticipated discoveries are made.

Libya currently produces 2 million barrels per day and may have the physical resources to produce more. The Libyan National Oil Company had projected 2.4 million barrels per day of production in 1980. Libya's light, low-sulfur oil is contained in highly porous reeflike structures, and is therefore both high quality and easy to produce. Yet the price increases of the past year may make the revenues resulting from production increases unnecessary. Production of 2 million barrels per day in 1985 and 1990 is anticipated.

Venezuela is undergoing a transition as an oil producer. Large thermal recovery projects are being introduced to its large fields to prevent production declines. These projects will allow continuous production of 2.2 million barrels per day from oil fields. Additions will be made to Venezuelan production for its continental shelf, where exploration is underway. Initial production from this source should be available by 1985, and significant contributions can be expected by 1990, yielding production estimates of 2.2 and 2.5 million barrels per day in these two years. Venezuela's significant heavy oil resources probably will not produce more than 150,000 barrels per day by 1990, owing to national policy decisions.

Low-absorber nations. Four OPEC countries—Kuwait, Qatar, Saudi Arabia, and the United Arab Emirates (UAE)—are designated as "low absorber" because their capacity to absorb oil revenues in productive investment is low. Qatar is the smallest, with production held to a government-imposed level of 500,000 barrels per day. While production could be higher, especially since onshore reserve estimates are expected to increase, government-imposed ceilings are likely to continue.

The UAE and Kuwait are mid-sized producers in this group, and both constrain production because of limited needs for revenues. The UAE is a coalition of several states, among them the producing regions of Abu Dhabi, Dubai, and Sharjah. The bulk of UAE production comes from Abu Dhabi, which has reserves of 30 to 40 billion barrels and current production of 1.5 million barrels per day. Some estimates of future Abu Dhabi production reach 3 million barrels per day, but at present the Abu Dhabi government is presumed to prefer conservation to more revenues than it can expend productively. Minimum estimates of production are 1.5 million barrels per day in 1985 and 1990. Total UAE production is projected to be 2 million barrels per day in 1985 and 1990.

Kuwait reserves of 69 billion barrels could support the existing production capacity of 3 million barrels per day. A country with a larger population and equivalent resources would probably produce 6 million to 7 million barrels per day, but for Kuwait production at this level would result in revenues for which no productive domestic use exists. Low revenue requirements should constrain Kuwaiti production to 1.6 million barrels per day through the 1980s, sufficient to finance the real equivalent of Kuwait's present budget, and extend its reserves for 100 years, a government goal.

Saudi Arabia. Saudi Arabian production is estimated at 8.5 million barrels per day in 1985 and 1990. Larger levels are sustainable, given the extent of Saudi reserves and a feasible rate of infrastructural development (including desalinization projects for pressure maintenance). However, the revenue requirements of the Saudi regime and the socially destabilizing consequences of rapid economic growth should prohibit an expansion of output from this desired level.

Saudi reserves are immense—the official figure of 166 billion barrels may understate their ultimate production. The largest fields are being produced conservatively, and the significant offshore potential in the Persian Gulf is still to be developed. Disallowing time constraints, the maximum feasible production from known Saudi fields (some of which are not in production at present) appears to be 18 million to 22 million barrels per day. A plan submitted to the Saudis by Aramco (the foreign managers of Saudi reserves) for modest capacity expansions in the early 1980s conforms to capacity of 12 million barrels per day in 1985. Yet it is doubtful that this level will be approved by the Saudis.

The Saudis will have goals other than revenues when deciding how much oil to produce. Over the past several years, they have repeatedly made clear the following goals:

- A resolution to the Arab-Israeli conflict that allows the Palestinians self-determination;
- A program to implement the goal of third world economic development as presented in the United Nations' New International Economic Order;
- A workable solution to the problem of investing surplus oil revenues over the long term.

The likelihood is low that the Saudis would set a production level that would be crippling to the Western economy. Saudi oil sales and much of their investment portfolio are tied to the dollar, so that economic injury to the United States would be felt directly by the Saudis. Furthermore, a weakened economy would compromise the West's ability or will to defend the Saudi government from either an internal or a Soviet threat. On the other hand, the Saudis have economic leverage and will probably use it to pursue the larger political goals listed above, particularly with regard to the Middle East. King Khalid and Crown Prince Fahd have stated their willingness, once a Palestinian solution is found, to produce "any level that is feasibly possible." While reports from Saudi Arabia depict a growing belief by technical experts that production should be curtailed, to do so would force major changes in the economic and political linkages between Saudi Arabia and the West. Barring unforeseen circumstances, this remains unlikely.

Domestic political considerations will also enter into the Saudi production decision. For example, expansion of oil-producing capacity would call for the introduction of more foreign workers, since the Saudi work force is not large enough to perform all the necessary labor itself. Such "guest workers," mostly from Yemen, Egypt, and the dispersed Palestinian community, are seen as a destabilizing force in Saudi society. Similarly, many Saudi technical experts, disgruntled with alleged corruption in the extensive royal family and with production levels in excess of Saudi Arabia's strictly defined current revenue

TABLE 1.8

Crude Oil Supply in Non-OPEC LDCs, 1978, 1985, and 1990
(in millions of barrels per day)

Area	1978	1985 (projected)	1990 (projected)
Africa			
Angola	0.1	0.3	0.3
Egypt	0.5	1.0	1.0
Tunisia	0.1	0.1	0.2
Other	0.1	0.4	0.8
Subtotal	0.8	1.8	2.3
Asia			
Australia	0.4	0.5	0.5
Brunei	0.2	0.3	0.3
India	0.2	0.4	0.4
Malaysia	0.2	0.4	0.3
Other	0.1	0.2	0.4
Subtotal	1.1	1.8	1.9
Latin America			
Argentina	0.4	0.5	0.6
Brazil	0.2	0.3	0.2
Mexico	1.2	4.5	5.5
Other	0.6	0.6	0.4
Subtotal	2.4	5.9	6.7
Middle East			
Oman	0.3	0.4	0.4
Syria	0.2	0.3	0.3
Other	0.1	0.1	0.1
Subtotal	0.6	0.8	0.8
Total	4.9	10.3	11.7

Source: Congressional Budget Office.

needs, are said to be arguing within the government for lower production levels. Finally, oil production is centered in the Shiite region of Saudi Arabia, a predominantly Sunni nation, leaving the oil fields a potential hostage in the event of domestic instability.

Yet political considerations argue for higher production levels as well. Reducing production would not lower revenues to producers, since oil prices probably would rise more than proportionately. Reduced production levels, moreover, could undermine the West economically, and damage the premise of the Western military umbrella. Many feel that as long as the Saudi royal family is in power, production below the current desired level of 8.5 million barrels per day is unlikely.

Supply in Non-OPEC Less-Developed Countries

Third world nations outside of OPEC are projected to produce 10.3 million barrels of oil per day in 1985 and 11.7 million in 1990. Country-by-country production estimates are given in Table 1.8. Outstanding among this group is Mexico, where production is estimated at 4.5 million barrels per day in 1985 and 5.5 million in 1990. Outside of Mexico, and perhaps Egypt, no country in this group can count on being a major long-term producer unless significant new discoveries occur. Argentina, the Sudan, and Vietnam are frequently cited as candidates for large new fields in virgin areas, but realistic estimates of their future production cannot include possible undiscovered resources.

Mexico's oil in the ground is given as 200 billion to 300 billion barrels, and the now official proved reserve estimate of 31 billion barrels is an understatement. Mexican planners are said to work with the figure of 60 to 80 billion barrels in ultimate production. Mexico is currently building pipelines for possible oil exports to Japan and an offshore mooring buoy for ultralarge crude carriers in the Atlantic. Both will be completed in a few years, giving Mexico the infrastructural capability to export its production.

PEMEX, the Mexican state-owned monopoly, has discussed plans for expanding output to 4 million barrels per day in 1982, and this could be fulfilled with little difficulty. Mexican officials are planning to drill 3,000 new wells by 1985. This would imply rapidly expanding production and dramatic increases in revenues; 4.5 million barrels per day in 1985 and 5.5 million in 1990 are anticipated. For this expansion to occur, however, infusions of foreign capital are needed, calling for additional borrowing by Mexico, a nation with significant foreign debts. Yet such borrowing may be allowed because Mexico's balance-of-payments deficit and unemployed population are growing at alarming rates.

TABLE 1.9

Projected Oil and Gas Balance in the Soviet Bloc and China, 1985 and 1990

Area	1985			1990		
	Oil[a]	Gas[b]	Total Oil Equivalent[c]	Oil[a]	Gas[b]	Total Oil Equivalent[c]
Soviet Union						
Production	10.1	51.6	19.4	9.6	65.9	21.5
Consumption	10.1	44.0	18.0	11.7	53.5	21.4
Net exports (+), imports (−)	+0.0	+7.6	+1.4	−2.1	+12.4	+0.1
Eastern Europe						
Production	0.4	6.7	1.6	0.5	6.7	1.7
Consumption	3.1	10.6	5.0	3.8	12.9	6.1
Net exports (+), imports (−)	−2.7	−3.9	−3.4	−3.3	−6.2	−4.4
Total Soviet Bloc net exports (+), imports (−)	−2.7	+3.7	−2.0	−5.4	+6.2	−4.3
People's Republic of China, net exports (+)	+1.0	0.0	+1.0	+2.5	0.0	+2.5
Total, Soviet Bloc and China			−1.0			−1.8

aMillions of barrels per day.
bBillions of cubic feet per day. A billion cubic feet is equivalent to 180,000 barrels of oil.
cMillions of barrels per day.
Source: Congressional Budget Office.

40

Egypt is the second largest producer in this group. Egypt's reserves are currently appraised at only 3 billion barrels, but a promising exploration program is underway in the Gulf of Suez and the Mediterranean off the Nile delta. Egypt also has reserves on the Sinai peninsula. Egyptian production will reach a level of 1 million barrels per day in 1985 and 1990.

The eastern continental shelf of South America has traditionally been considered a promising geological region, particularly offshore Argentina. Several onshore producing areas in Argentina are found in sedimentary basins that extend onto the shelf. The Argentinian national oil company is for the first time inviting foreign firms onto the shelf for joint exploration. Discoveries thought possible by some geologists could bring Argentinian production to over 1 million barrels per day by the late 1980s; in their absence, however, onshore production will sustain a level of 0.5 million barrels per day in 1985 and 0.6 million in 1990.

Similar optimism was held for the Brazilian shelf, but except for the Campos Basin it has not proved out. Petrobrás, the state-owned company, is conducting a last sweep of the shelf and will finish its preliminary exploration in the early 1980s. Future discoveries are thought possible in the western Amazon region, but logistical difficulties may make production extremely difficult.

India has also had recent discoveries, and new fields are thought possible in the Bay of Bengal. Australia is conducting an exploratory program that is thought likely to be successful. Tunisia and Malaysia are likely to expand production and are seen as promising areas for exploration.

Finally, some areas are viewed as being exceptionally promising, but have not been explored adequately and are not yet producing. Most notable are the Sudan and Vietnam, where political constraints have limited activity. Cameroon, Chad, Guatemala, and Pakistan are also cited as prospective producers.

Supply in the Soviet Bloc and China

While Chinese oil production is projected to expand in this decade, declines in Soviet production should move the Soviet bloc into a net importer position by 1985. The Soviet bloc may draw almost 2 million barrels per day from the world market by 1990. Table 1.9 presents estimates of the net energy balance for the Soviet bloc and China in 1985 and 1990, respectively. Production and reserve data for the Soviet Union, Eastern Europe, and China are relatively difficult to obtain and are not always consistent but suffice to allow estimates of future production. For the purposes of projecting the state of the international oil market, however, the net oil exports of these countries

must be estimated. Soviet bloc exports are currently positive, and
should remain so through the early 1980s, when production from
Soviet fields will decline for a peak expected in 1980-82. Expansion
of Chinese oil supply, however, will compensate for some of the decline in Soviet and Eastern European net exports.

Soviet Union. Soviet crude oil reserves are now set at 71 billion
barrels, although some estimates of Soviet ultimate production exceed
125 billion. This is the world's second largest proved reserve. Soviet
gas deposits are even larger. The Soviet Union's 910 trillion cubic
feet of gas exceed the gas reserves of the entire Middle East.

Several critical questions dominate analysis of Soviet production.
The first is the status of Samotlor, the largest Soviet field. Samotlor
is said to be experiencing declining production and to be headed toward
depletion, making it reasonable to predict that Soviet production is
nearing a peak. A second question concerns reservoir mismanagement,
specifically improper waterflooding leading to losses in ultimate recovery. Older Soviet fields in the Ural-Volga region are reported to
be pumping 50 to 60 percent water. Newer fields in the western
Siberian region, however, apparently do not have this problem.

The Soviets appear to be concentrating on drilling more wells
in the already producing western Siberian fields, choosing the more
certain activity of extending known reserves over exploration in the
Kara Sea and Caspian Sea although the rate of discovery there is high.
Thus, the long-term prospects for Soviet production could be good.
Despite predictions to the contrary, and difficult drilling conditions,
the Soviet Union has been able to continue increasing production and
has pledged to deliver oil and gas to their export customers in the
1980s.

The withdrawal of Western technology following the Soviet invasion of Afghanistan may make it difficult for the Soviets to fulfill production plans, at least in the near term. Because their drilling capabilities are behind those of the West, the Soviets had sought an infusion
of Western know-how in this area. Since many of the new oil and gas
fields are inhospitable areas inside the Arctic Circle or in Siberia,
and are also at great depths, this technological assistance was considered critical in maintaining Soviet production. In its absence, there
is little doubt that production will fall in a few years.

How the Soviet shortfall will be distributed is unclear. On the
one hand, reduced gas deliveries to Eastern Europe would threaten
the political unity of the Soviet bloc. On the other hand, gas would be
an important source of cash for the Soviet Union if it was sold in increasing amounts to Western Europe. Pipelines have already been
built toward this goal. Moreover, if a period of political tension between the United States and the Soviet Union is at hand, the Soviets
may seek to build stronger economic ties to Western Europe to weaken

its alliance with the United States. Thus, much of the shortfall of oil
and gas production may be absorbed inside the Soviet bloc through
reduced consumption and slower economic growth, rather than by the
world market. Because the Soviet Union's economy is centrally planned,
the government can specify energy use. The 1976-80 economic plan
curtailed auto production dramatically, and the government is expanding electrification through nuclear power.

Eastern Europe. Eastern European countries have traditionally
relied on the Soviet Union for energy imports. Soviet pipelines to
Eastern Europe have already been constructed to carry future imports
of Soviet natural gas. These countries are, however, negotiating with
Middle Eastern and North African producers for crude oil, and are
building refineries that will handle the specific characteristics of
these crudes. Eastern European oil production is limited to Rumania
and Yugoslavia, and prospects for its expansion are limited. Some
natural gas is being found, and its production will probably increase.
Total required Eastern European oil imports will rise, however, from
their current 2 million barrels per day to 3.3 million barrels per day
in 1990 (see Table 1.9).

People's Republic of China. Over the past few years, Chinese
reserve figures have escalated substantially. China's onshore resources
are said to be 40 billion barrels, and its offshore potential is possibly
as large, in contrast to the government estimate of 20 billion barrels.

Given China's new plans for development and its new access to
Western technology, energy consumption and production should both
increase rapidly. Estimates of 1985 Chinese oil production have been
in the range of 4.1 to 6.1 million barrels per day. The upper end of
this range will be difficult to achieve, however, because of an absence
of infrastructure such as port facilities. Production of about 4 million
barrels per day is more likely, including about 2 million barrels per
day from the large northeastern fields, 1 million to 2 million barrels
per day from the western fields where development is now planned,
and about 0.5 million barrels per day from offshore Po Hai Bay.
Other offshore formations, such as the Taiwan Straits and South
China Sea, may be in production by 1985, and will probably be producing significant quantities by 1990.

Production of 4 million barrels per day in 1985 and 6.5 million
in 1990 would create an exportable surplus of 1 million barrels per
day in 1985 and 2.5 million in 1990. But these export levels would not
balance the net imports of the Soviet Union and Eastern Europe.

Oil Prices in the 1980s

World demand for oil, including U.S. net demand for imports,
will exceed available supply by 4.4 million barrels per day in 1985

TABLE 1.10

Projected World Oil Balance, 1985 and 1990
(in millions of barrels per day)

	1985			1990		
	Production	Consumption	Imports (−) Exports (+)	Production	Consumption	Imports (−) Exports (+)
Industrialized countries						
United States	9.4	19.5	−10.1	8.5	19.9	−11.4
Canada	1.5	2.0	−0.5	1.9	2.2	−0.3
Japan	–	6.5	−6.5	–	7.4	−7.4
Western Europe	4.2	15.9	−11.7	4.2	17.3	−13.1
Subtotal, industrialized countries	15.1	43.9	−28.8	14.6	46.8	−32.2
LDCs						
OPEC	30.1	4.3	+25.8	31.3	6.3	+25.0
Non-OPEC	10.3	10.7	−0.4	11.7	13.3	−1.6
Subtotal, LDCs	40.4	15.0	+25.4	43.0	19.6	+23.4
Subtotal, non-Communist world	55.5	58.9	−3.4	57.6	66.4	−8.8
Soviet Bloc and China, net imports			−1.0			−1.8
Total imports (−)			−4.4			−10.6

Source: Projected by the Congressional Budget Office, using price assumption of $30 per barrel in the fourth quarter of 1979, rising at an annual rate 2 percent greater than the general price level.

and by 10.6 million barrels in 1990 at the level of prices used for this analysis (see Table 1.10). Thus, it is possible that the world oil market will remain tight through the next decade, and there will be increases in the price of oil.

Should the conditions just described prevail, excess demand will probably bring price increases for oil greater than used for these projections by the mid-1980s. Theoretically, price increases should ultimately cease when oil prices become equal to the prices of synthetic substitutes. However, this is true only for the long run. The time required to bring about the production of synthetic substitutes makes this constraint ineffective in the short run. Moreover, price increases tend to be reinforced by the willingness of consumers and refiners to pay them, up to the point at which the high prices have negative effects on the economies of industrialized countries. By slowing down economic growth and reducing consumers' disposable income, higher prices ultimately soften the demand for oil and restabilize the market. This happened to some extent in 1975, and in 1980.

Even if supply and demand were projected to be in balance throughout the coming decade, higher oil prices would remain a possibility. This could happen if producers decide to restrict supply and raise prices. Many producing countries perceive that their rate of economic growth has reached a practical limit because of the inflationary or socially destabilizing effects faster growth would have. They may conclude that if revenues cannot be translated effectively into growth, oil is best left in the ground. When market prices are rising, a producer in this situation may choose to reduce output rather than increase it, thus earning the same revenue on less production. Here again, the effects of higher prices on the economies of the industrialized countries would act ultimately to restrain such increases. Before that point is reached, however, price increases are possible even in the face of stable demand.

Price increases might take a hidden form. If OPEC decided to denominate oil rates in currencies other than the dollar, this would raise the price of oil paid by many other industrialized countries in the future since their currencies tend to appreciate relative to the dollar. In addition, U.S. import prices would rise by the extent to which the dollar fell in value relative to the currency in which oil was priced. Hidden price increases might also occur if oil were transferred away from contract sales to spot sales. Spot prices will remain higher than contract prices as long as availability is considered uncertain. In such a situation, refiners would pay a premium that reflects the cost of closing their refineries.

Finally, hidden price increases could take the form of economic concessions to producers. Many producing nations are building refining and petrochemical facilities that will come on line in the mid-1980s.

Because of the difference in transportation costs in shipping crude and refined products (crude, being dense, is cheaper to ship), and because of projected excess capacity, these refining and petrochemical facilities may not be economically viable in a conventional market. Thus, producers may tie sales of these products to purchases of crude oil. Such "tie-in" sales would involve an additional transfer of resources to OPEC countries and would amount to a de facto price increase. A similar disguised resource transfer might occur in the form of trade concessions offered to producing countries by consumer countries competing for access to oil. Bilateral agreements between producing and consuming countries are becoming increasingly common as consumers seek to secure oil deliveries and promote reciprocal trade with producers. These agreements not only often raise prices directly through "goodwill payments" and the like, but may also involve financial or trade concessions to producers. This type of hidden price increase is often in the interest of both parties since it can lead to more certain deliveries and better trade balances.

Effects of Higher Prices on U.S. Economy

The likelihood of higher oil prices in the coming decade raises the question of how these price increases will affect the U.S. economy. Higher oil prices in the 1970s contributed to the simultaneous existence of inflation and unemployment.

The effects of higher oil prices on inflation are obvious. In 1980, the United States will consume approximately 6.5 billion barrels of oil. A price increase of $4 per barrel would raise the rate of inflation by one percentage point. Price increases in substitutable fuels, such as coal and natural gas, would exacerbate this inflationary effect.

On the basis of the price assumption made for this analysis ($30 per barrel in the fourth quarter of 1979, increasing at 2 percent annually in real terms), oil prices may rise to $52 per barrel by 1985. This may, as already noted, be a minimum figure, since an excess demand for oil of approximately 4.4 million barrels per day is projected for that year. Assuming that oil prices rise by a minimum of $20 per barrel over the six-year period and projecting oil consumption in 1985 at 7.1 billion barrels (19.5 million barrels per day), price increases between now and 1985 could cost consumers $142 billion in that year. This would amount to a cumulative increase in the general price level of 3 percent annually between 1980 and 1985, not taking account of price increases in other fuels, cost-of-living wage increases, and other costs related to the price of oil.

Rising oil prices not only increase the rate of inflation but also dampen economic growth. Accompanied by higher prices for other sources of energy, they reduce the disposable income of consumers

and hence the aggregate demand for goods and services. This demand
reduction has a depressing effect on the national economy. Higher
energy prices also lower business profits, reducing the volume of
new investment. The inflationary effects of higher energy prices
serve to lower the real interest rate, making saving less attractive.
The expectation of continued increases in energy prices may further
reduce incentives for investment, causing investors to turn to speculative assets such as gold. All of these effects diminish capital formation, and thus economic growth and productivity.

Major Sources of Uncertainty

Any estimate of future energy supply and demand is subject to
a number of major uncertainties.

Levels of economic growth. The levels of crude oil demand
presented in this study are highly sensitive to changes in economic
growth. A one percentage point annual increase in the worldwide rate
of economic growth would increase crude oil demand by 4 million to
5 million barrels per day in 1990. A comparable decrease would lower
demand by the same amount. It may be that the rates of growth assumed here are optimistic. This would close the gap between projected
demand and supply, but at the cost of protracted slow growth. Shifts
in demand may occur among the less-developed countries, in which
the greatest uncertainty as to future growth levels exists. This relationship suggests that surges in economic growth in the coming decade
will be self-adjusting; as higher worldwide economic growth leads to
higher oil prices, the higher prices in turn will inhibit that growth.

Disruption of supply. The supply estimates presented previously
presume that no political or logistical event interrupts supplies from
producing nations. Recent experience, however, shows that oil supplies can be disrupted either intentionally or accidentally in a number
of ways. Future disruptions, beyond limiting available supplies, would
serve to reallocate supply within the OPEC cartel. In the past, supply
disruptions have often served as the starting point for price increases.

Production sharing. While OPEC has never formally allocated
world production among its members, the possibility cannot be over-
ruled. It would be most likely at a time of slack demand and falling
oil prices. The cartel might then seek to restrict production to predetermined levels.

Efficiencies in utilization. The main uses of oil are in transportation and space heating. The demand projections presented here assume that there will be no unforeseen technological innovations in
those fields. Significant advances in automotive engineering, archi-
tectural design, or industrial process design would fundamentally
change the outlook for the world oil market. As the time horizon of

these estimates recedes, energy use patterns become an increasingly important source of uncertainty. Demand by the industrial nations is projected to reach 46.8 million barrels per day in 1990 (see Table 1.4). Thus, a 20 percent increase in the overall energy efficiency of the industrial economies over what is assumed in the forecasts would eliminate the excess demand for oil projected for that year. Such an increase cannot be anticipated, but neither can it be ruled out.

Production from non-OPEC sources. Uncertainties also exist as to levels of future production in many non-OPEC countries and the Soviet Union. While there is reason to believe that nearly all of the world's giant oil fields have been discovered, many less-developed countries have hydrocarbon prospects that are underexplored. Recent exploratory successes in India, Pakistan, and Vietnam, for example, suggest potential additions to world reserves. Given the low level of geological knowledge in many of these areas, unforeseen discoveries are possible. Nevertheless, the time necessary to bring these discoveries to production would rule out a contribution from this source until the late 1980s. Moreover, many LDCs that discover oil would increase their domestic consumption rather than export any new production.

Information on Soviet production is often contradictory. Most analyses depict Soviet production as approaching a peak in the early 1980s. Some observers, however, believe that Soviet resources are currently underproduced, and that the reserve base will be expanded. This uncertainty is compounded by the uncertainties of U.S.-Soviet relations and how they will affect Soviet oil and gas production.

Alternative fuels or energy sources. The demand for oil could conceivably be reduced by an accelerated development of alternative fuels. Most notably, increases in the availability of natural gas might occur, partly as a result of the decontrol of the U.S. gas market. The development of alternative energy sources, particularly of solar energy or synthetic liquid fuels, might erode the demand for oil by the late 1980s. The future of nuclear power is also uncertain. Accelerated nuclear programs, including greatly expedited licensing procedures, could erode world oil demand by approximately 2 million barrels per day by 1990.

SECURITY OF SUPPLY

The security of the U.S. oil supply has been a major concern of energy policy since the OPEC embargo of 1973-74 threatened Western and Japanese security of access to a critical source of energy. Despite the risks associated with importing foreign oil, particularly from the Middle East, total U.S. oil imports, as well as imports from the Middle East, have increased significantly since then.

During this period, however, certain measures have been taken that could deter another supply interruption or reduce its impact. The International Energy Agency (IEA), founded at the initiative of the United States following the 1973-74 oil crisis, is intended to handle such a supply interruption. Through the IEA, an oil-sharing mechanism has been worked out among the member countries. In the event of an interruption, members will undertake predetermined emergency conservation measures, and available oil supplies will be allocated among them. Every IEA participant is to set aside an emergency reserve equal to 60 days of oil imports.

The Strategic Petroleum Reserve

In the Energy Policy and Conservation Act of December 1975 (EPCA), Congress mandated the development of a Strategic Petroleum Reserve aimed specifically at offsetting the likely output losses from a foreign oil supply interruption. To expedite the development of the reserve, the legislation provided for an early storage reserve of at least 150 million barrels. It also required the Federal Energy Administration to submit a plan for designing, constructing, and filling the Strategic Petroleum Reserve. The plan was submitted to Congress in February 1977 and became effective in April. It superseded the Early Storage Reserve plan but retained the goal of storing 150 million barrels of oil by December 1978.

In May 1977, the plan was amended to accelerate the Strategic Petroleum Reserve by placing 250 million barrels in storage by December 1978, another 250 million by December 1980, and an additional 500 million barrels by 1985. This would bring the total crude in storage to 1 billion barrels—the maximum amount authorized by EPCA. Currently about 92 million barrels are in storage, with capacity for an additional 156 million barrels available. Construction of an additional 290 million barrels of storage capacity, which would bring total capacity to 538 million barrels, is scheduled for completion in 1982. No specific plans have been developed to bring capacity up to the 1 billion barrel goal. No oil purchases have been made since the spring of 1979. The Department of Energy's plans for oil purchase are presently unclear. The fiscal year 1981 budget request called for the resumption of oil acquisition in June 1980, at the rate of 3 million barrels per month, increasing to 7.5 million barrels per month in 1982. The revised budget request submitted in March 1980, however, delays purchases until June 1981. Recent statements of DOE officials indicate that no final decision regarding purchase has been made at this time.

Emergency Distribution Measures

During the 1973-74 crisis, Congress passed the Emergency Petroleum Allocation Act of 1973, providing for a system of petroleum allocation regulations. Essentially, the allocation system gave priority to the protection of public health, safety, and welfare and to national defense, mineral production, and agriculture. Since then, the Federal Energy Administration and the Department of Energy have conducted a careful review of the policies pursued during the 1973 oil embargo. The department has funded a number of studies to evaluate the existing priority classification system, and has continually updated the allocation system used during the Iranian disruption of 1979, even though the effectiveness of that system was questionable.

In 1979, the Emergency Energy Conservation Act authorized the president to develop a standby gasoline rationing plan to be implemented at his discretion when there is a 20 percent drop in gasoline availability. The rationing plan would be subject to congressional approval, and either house could veto a presidential decision to implement it. The act also mandated a gasoline energy conservation plan, to be implemented by the federal government and the states, although many states have not yet developed their parts of the plan as required by the act.

Factors Affecting the Security of Supply

The security of the U.S. oil supply is dependent upon a number of political and economic factors. These include the size and sources of U.S. imports, the tightness of the world oil market, the size of the OPEC surplus, and the political stability of the oil-exporting countries and of U.S. relations with them.

Size and Geographical Concentration of Oil Imports

Since the 1973 oil embargo, the United States has dramatically increased its oil imports. Net refined and crude imports rose from about 6 million barrels a day in 1973—representing 36 percent of U.S. oil consumption—to about 8 million barrels a day in 1979—approximately 44 percent of U.S. oil consumption.

The countries responsible for the 1973 oil supply interruption now provide a much larger share of U.S. petroleum imports. In January-September 1973, the Organization of Arab Petroleum Exporting Countries (OAPEC) provided 23.6 percent of total U.S. imports. By 1978 its share had risen to 39 percent, or 17 percent of U.S. domestic demand. The states belonging to the larger Organization of

Petroleum Exporting Countries (OPEC) increased their share of U.S. oil imports over the same period from 70 to 80 percent.

As pointed out previously, the size and importance of U.S. imports will continue to increase over the coming decade. From 44 percent of U.S. demand in 1979, oil imports are projected to rise to 52 percent of demand in 1985 and 57 percent in 1990. The U.S. economy will thus become increasingly dependent on foreign oil.

Since the 1973 quadrupling of oil prices, exploration and development have been intensified in various parts of the globe, such as the North Sea, Mexico, China, and West Africa. This should lead to a greater diversification of import sources. If Mexico, for example, has the proved reserves indicated previously, the U.S. dependence on OPEC could be somewhat reduced in the near future. An oil supply interruption is much less likely in Mexico than in the Persian Gulf. Yet, the supply projections indicate that new producers such as Mexico will not bring a significant change in the geographical distribution of U.S. imports. The bulk of U.S. oil imports will continue to come from Persian Gulf and North African sources.

Tightness of World Oil Supplies

Security of supply is also dependent on the conditions of world oil supply and demand. As world oil markets tighten, the industrial oil-importing countries become more vulnerable to a supply interruption, and oil exporters have more leverage. If an embargo were to occur when world oil supplies were tight, exporters not involved would not be likely to have much excess capacity. In 1973-74, the United States was able to pick up enough excess supplies to mitigate the effects of the interruption. In the early 1980s, the world oil market could have some slack, in the absence of a crisis similar to that of Iran. But this slack should disappear by 1985, along with the development of strong upward price pressures. Tight markets appear likely throughout the second half of the decade, and few countries would be capable of providing short-term production increases if a disruption occurred.

The OPEC Capital Surplus

The financial reserves of many OPEC countries skyrocketed during the 1970s. Many of them have engaged in ambitious development plans, and have made large spending commitments. An interruption of any length in oil shipments could disrupt or suspend these development plans, with adverse domestic political consequences. For this reason, the potential participants in a new interruption would be among the low-absorbing nations—those countries that have difficulty finding economic uses for all of their oil revenues.[3] Libya,

Saudi Arabia, and Kuwait are typical low-absorbing countries. Nigeria, on the other hand, is an example of a high-absorbing country that needs oil revenues to sustain its development plans. While the Iranian revolution of 1979 may cause some OPEC countries to reevaluate their development plans in the short run, most of the high absorbers are likely to continue to spend the bulk of their oil revenues.

Some of the OPEC countries not only have heavy domestic investments but have also acquired extensive dollar-denominated assets abroad. A crippling of the Western economy would clearly be disadvantageous for them. The Western industrial countries also provide the market for new exports sought by OPEC members as part of their industrial diversification programs.

The OPEC surplus, which had been declining in 1977 and 1978, was dramatically increased by the price rises of 1979. These new revenues may be so large that they cannot be absorbed efficiently into OPEC economies. The OPEC countries lack a developed commercial infrastructure. In short, there is a limit to the amount of new investment that can be accommodated, given the existing transportation system, limited supplies of skilled labor, and the small potential of the domestic market. There are also drawbacks to the investment of OPEC's surplus abroad. Purchases of many assets in industrial countries are restricted, and the U.S. seizure of Iranian assets in 1979 may have provided an additional incentive to hold the surplus in more liquid form in the future. OPEC purchases of gold may have been an important factor in the upsurge of gold prices in early 1980. On the other hand, even the most liquid assets cannot always be converted into real purchasing power in time of crisis. If a situation like that of 1973-74 were to recur, the nature of the emergency would be likely to prevent the OPEC countries from converting their assets into goods. The international gold market, for example, might be severely restricted under such circumstances.

Political and Logistical Factors

Continuity of oil supplies also depends upon political factors. World political conditions change rapidly, making it difficult to assess the likelihood of another supply interruption. A serious interruption could occur in the even of civil war, revolution, or serious domestic strife in a major oil-exporting country. A prime example is the Iranian revolution. While other producers offset somewhat the slackening of Iranian supply in 1979, they would be less likely to make up the difference in the 1980s when significant excess capacity will not exist.

International tensions could also lead to an interruption, even at some sacrifice to the producers involved. A radical government might be willing to absorb economic losses to achieve ideological goals. Perhaps the most likely cause of such a supply interruption would be another Arab-Israeli war. Each of the four episodes of open warfare between Arabs and Israelis has led to some disruption in the world oil trade. The 1948 confrontation led to the permanent closing of the Iraq Petroleum Company's pipeline to Haifa. In 1956 the passage through Suez was interrupted—the major oil route at that time from the Middle East to the West. From the 1967 Yom Kippur War until 1975, the Suez Canal remained closed. The 1973 crisis led to the OPEC embargo. While the initiatives of the Carter administration in the late 1970s seemed to have reduced the prospects for warfare in the Middle East, the Soviet invasion of Afghanistan in 1980 once again raised the specter of a break in supplies from this region.

An interruption could be caused by logistical breakdown. Logistical breakdowns include explosions in oil pipelines and oil field fires. Iraq experienced a pipeline explosion (possibly caused by terrorists) in 1978. A fire in eastern Turkey knocked out Mediterranean delivery of Iraqi oil blend for two weeks in 1978, amounting to 500,000 barrels a day. Similarly, a 1977 fire at a Saudi Arabian producing facility resulted in a loss of world oil supplies equivalent to a shutdown of the Trans-Alaskan Pipeline (1.5 million barrels per day).

A number of factors work against another interruption in the near future. These include the development plans of many OPEC countries, their heavy investment in the United States, the chance that the United States could switch to non-OPEC oil for the short run, and the difficulties in effectively implementing an interruption. Nonetheless, as long as the United States remains heavily dependent on foreign oil, particularly from the Middle East, the security of its oil supply is an appropriate concern of U.S. energy policy.

Economic Effects of an Oil Interruption

If another interruption in the supply of oil were to occur, what would be its consequences for the United States? The following section gives estimates of the impact on the U.S. economy of a one-year interruption in 1984 and 1990.[4]

Impact of a 3.5 Million Barrel per Day Interruption in 1984

While the 1973-74 oil crisis lasted for about five months, in this analysis the interruption is assumed to last one year. This assumption makes it possible to estimate the maximum impact the interruption would have on the economy. The petroleum shortfall is assumed to be 3.5 million barrels a day, representing approximately

8 percent of projected energy consumption in 1984. The size of the shortfall is aimed to fit a wide range of possible events likely to lead to a supply interruption, such as an Arab-Israeli war or a shutdown of a major producer because of internal political problems. Oil prices are assumed to increase by 20 percent from their preinterruption levels. The U.S. Strategic Petroleum Reserve is assumed to contain 500 million barrels and to be depleted after one year. Oil allocation regulations are also assumed to be in effect, and are maintained throughout the one-year period.

Price controls on oil and oil-related products are assumed to be in effect throughout the crisis. It is quite clear that without controls oil price increases combined with shortages in those industries that use petroleum as inputs could lead to short-run economic problems. Consequently, it is assumed that, during an emergency, the U.S. government would implement some form of price controls. Finally, the analysis also assumes that conservation measures as well as fuel switching are implemented as far as possible.

A 3.5 million barrel per day interruption (which represents a 9 percent petroleum shortfall) would have a significant impact on the American economy (see Table 1.11). Real output would decline by 6.6 percent (a drop of $272 billion in 1984), while the unemployment and inflation rates would increase by 2.1 and 20.0 percentage points, respectively.[5] These price increases would also have an impact on GNP: higher prices reduce both real income and real wealth, causing households to reduce their purchases of goods and services, thereby slowing economic growth.[6] A number of additional simulations were also calculated, which indicated that larger petroleum shortfalls would lead to proportionately larger output losses, greater unemployment, and more rapidly rising prices.

In the three-year period following the oil production limitation, the economy would rebound substantially as real output increased, unemployment declined, and prices subsided. But real GNP would still be below the baseline forecast, and the unemployment and inflation rates would be higher.

Impact of a 4 Million Barrel per Day Interruption in 1990

CBO projects U.S. imports to grow from 10.1 million barrels per day in 1985 to 11.3 million barrels per day in 1990 if oil prices increase at 2 percent annually in real terms. Imports will thus become a more significant percentage of U.S. oil and energy consumption. Growing U.S. dependence on foreign oil will obviously increase U.S. vulnerability to supply interruptions. The world oil market is expected to tighten in the late 1980s and early 1990s. Should an interruption occur during this period, the United States would not be able

TABLE 1.11

Impact of a Yearlong Oil Supply Interruption Amounting
to 3.5 Million Barrels per Day in 1984 on GNP,
Unemployment, and Inflation in 1984 and 1987

	1984	1987
Change in real GNP (in billions of 1984 dollars)	-272	-100
Change in real GNP (percent)	-6.6	-2.2
Change in unemployment rate (percentage points)	2.1	1.2
Change in inflation rate (percentage points)	20.0	2.5

Note: The effects are compared with CBO's economic
projection without a supply interruption.
Source: Congressional Budget Office.

to obtain significant oil supplies from countries not involved in the
supply interruption as it did during 1973-74. On the other hand, U.S.
oil supplies may become more diversified in the late 1980s when
Mexico, China, and possibly other new producers may become alternative sources of oil.

In an attempt to make some quantitative assessments of the
effects of a supply interruption in 1990, the results of the 1984 simulation were extrapolated. The 1984 interruption of 3.5 million barrels
per day, or approximately 35 percent of U.S. oil imports, is equivalent to about 4 million barrels per day in 1990. A supply interruption
of this size lasting for one year would cause real GNP in 1990 to drop
about 7.3 percent below the projection, and unemployment and inflation to rise by 2.4 and 22.4 percentage points, respectively. The
Strategic Petroleum Reserve required to prevent the GNP losses
would be approximately 1.5 billion barrels. Thus, over the long term,
the vulnerability of the United States to another supply interruption
would grow with increasing import levels.

Security of Supply and U.S. Foreign Policy

The Arab oil supply interruption in the fall of 1973 signaled that the United States had entered a new era of economic interdependence with other countries. This meant that the United States would, in the future, be increasingly vulnerable to foreign attempts to use economic leverage for political purposes. The insecurity of oil supply has had important implications for national security and U.S. foreign policy.[7]

Oil and the Western Alliance

The months following the Arab oil supply interruption were among the most turbulent in the 25-year history of the alliance between the United States and Western Europe and Japan, the nadir being the denial of U.S. access to European bases during the 1973 Arab-Israeli War. At the commencement of the 1973 embargo, U.S. strategy assumed that a unified opposition of oil-consuming countries, coupled with the threat of economic or military retaliation, would undermine the solidarity of the OPEC countries. At the Washington Energy Conference in February 1974, however, the major consuming countries failed to develop a common strategy. Despite repeated attempts by the United States to reach a compromise, the Europeans, led by France, resisted bloc strategies and sought instead to make bilateral political and economic arrangements with OPEC countries. By late 1975, at the Conference on International Economics, the United States was forced to accept that its policy was a failure and that OPEC could not be undermined. Some analysts attributed the failure to inevitable differences between the United States, an international power with global responsibilities and commitments, and the Europeans and the Japanese with their more regional orientation. For the United States, energy was not a prime factor in foreign policy, while the freedom to act in foreign affairs was clearly such. The Japanese and Europeans, on the other hand, were quite willing to merge their foreign policies and their energy policies.

Other strains have developed from international competition for oil in a tight market. Competition has been spurred by the ventures of national oil companies and the willingness of their governments to encourage and often subsidize direct state-to-state deals with OPEC members. By the end of 1979, for example, Japan had agreed to purchase oil from Iran at $40 per barrel, $17 above the world price, at a time of considerable tension between the United States and Iran. Such actions lead to increased costs for all oil consumers, and are a divisive force within the Western alliance.

Oil and the Soviet Union

The impact of the energy crisis on American–Soviet relations is a controversial topic. In one view, the U.S. oil crisis has until now only marginally or indirectly affected American–Soviet relations.[8] Yet, this view also holds that, if the United States cannot solve its long–run energy problems, or if there is a return to the cold war, the Soviet Union may be able to exploit this dependency. A second school of thought maintains that the energy crisis has already led to a decline in U.S. power and influence vis-à-vis the Soviets.[9] This group of analysts believes that, because the Soviet Union is currently self–sufficient in oil, it has an "extra card to play" in the world power struggle.

The first view. In the short run, according to the first school of thought, U.S. dependency on foreign oil is not likely to be one of the major factors determining the future course of Soviet–American relations. These analysts perceive the major U.S.–Soviet concerns as: first, the avoidance of armed conflict through the Strategic Arms Limitation Talks and détente; second, relations between China and the Soviet Union; third, the future of the U.S.–Japanese mutual security treaty; fourth, the prevention of Soviet hegemony over Western Europe; and fifth, the Middle East conflict between Israel and the Arabs. In the latter areas, there is a possibility of confrontation between the United States and the USSR, a confrontation that could be linked to the problem of energy. The intrusion of the Soviet Union into the area adjacent to the Persian Gulf has made the Persian Gulf a scene of potential U.S.–Soviet conflict. Yet many believe that the Soviet Union would gain little from a military intervention that closed oil supply lines.

If the United States continues to remain dependent on oil imports from the Middle East, the Soviet Union may come to view the U.S. oil dependency as something it can exploit in two ways. First, the USSR could play a more active role in supporting OPEC by providing more political, military, and economic support to the oil–producing countries. Second, it could sponsor political movement aimed at replacing the conservative monarchies of the Persian Gulf with more radical governments hostile to the West, although existing radical governments, such as those of Iran and Iraq, have not appeared to be avowedly pro–Soviet. Alternatively, it could exploit ethnic or religious division in Iran or Saudi Arabia.

The second view. Others hold that, as a result of the OPEC price hike and the continued dependence on foreign oil, the West, particularly the United States, has suffered a decline in economic and political power relative to the Soviet Union. Since the Soviet Union is still self–sufficient in energy, it now has an extra card to play in the

East-West struggle. This self-sufficiency, however, is projected to erode in the early 1980s, as Soviet production declines from its peak levels.[10]

Oil and the Middle East

The changes that have occurred in U.S. policy in the Middle East since 1973 are rather clear. Prior to the 1970s, the oil-consuming countries dominated the oil-producing countries of the Middle East. This domination rested on the major oil companies and on British or American influence in the oil-producing countries. It enabled the consuming nations to bring the producing countries into the world economy at terms highly favorable to the former. This domination was gradually eroded by the rise of Arab nationalism, the growth of Soviet power and influence, and the West's support of Israel. Simultaneously, most Western countries, such as the United States, became increasingly dependent on supplies of Middle Eastern oil.[11]

During this period, the United States has sought to keep the issues of the Arab-Israeli conflict isolated from its relations with the key oil-producing countries in the Persian Gulf (for example, Saudi Arabia). It is not certain, however, that these issues can be separated. Continued enmity between Israel and the Arab states has created obvious liabilities for the United States. Radical political movements in oil-producing nations are given political capital by the failure of Israel and its neighbors to reach agreement. This jeopardizes the stability of both these nations and their oil exports. Moreover, the regional response to the Soviet invasion of Afghanistan may have been tempered by Islamic nations' desires to distance themselves from the United States because of its support for Israel. This conflict between the commitment to Israel and its desire to stabilize its relations with oil producers reflects the impact of oil import dependence on U.S. foreign policy.

POLICY OPTIONS

Oil Imports in the 1980s

Because of anticipated declines in domestic production and moderate increases in demand over the next decade, U.S. oil imports are projected to grow from their current level of approximately 8 million barrels per day to 11.3 million barrels per day in 1990—barring policy changes or higher prices than those used for this analysis. The increasing imports will pose four distinct economic and political risks for the United States.

First, the geographical location of the oil supplies will make them vulnerable to disruption. Although new supplies will be forthcoming from countries such as Mexico and China in the next decade, more than half of the world's oil will still come from the Middle East and North Africa. It will be vulnerable to the same types of political and logistical disruptions that have been experienced over the past six years. Moreover, increased revenues in the Persian Gulf states may strain the social and political stability of those countries. In addition, anticipated declines in Soviet oil production raise the question whether the Persian Gulf may become an arena of conflict between the Soviet Union and the United States.

Second, oil price increases above those assumed in these projections appear inevitable in the 1980s. A recession in the industrialized countries over the next several years may soften the oil market in the short term. Yet, presuming an economic recovery by 1985, this slack will disappear, and price increases seem likely. By 1990, with only the price increases assumed in this analysis (2 percent in real terms annually), an excess demand of slightly over 10 million barrels per day will develop. The price increases of 1979 were caused, in large part, by a temporary excess demand far less than this amount.

Third, larger import volumes and higher oil prices will mean larger dollar outflows to producing nations. The capability of producing nations to absorb these revenues through development expenditures will be taxed severely, resulting in a larger "petrodollar surplus." This suggests that a smaller proportion of petrodollars will be recycled through U.S. exports to producing nations than at present, and that a larger share will seek assets in the Eurodollar market, in other consuming nations, or in speculative investments such as gold. The larger surplus may add to the volatility of the world monetary system, and induce producers to seek a currency or group of currencies other than the dollar in which to price oil. If OPEC were to abandon the dollar, it might signal the end of the dollar's use as an international reserve and transactions currency, and result in heavy losses to the U.S. economy.

Finally, an inexorably tightening oil market may have a profound effect on international relations. Competition within the industrialized countries for scarce supplies could lead to the acceptance of "hidden" price increases in the form of trade, aid, or other concessions made to producing countries on a state-to-state basis. Further declines in the value of the dollar, fueled in part by oil imports, could create tensions between the United States and its major trading partners. If, as seems likely, the Soviet Union becomes a net oil importer by 1985, it may participate in the competition for access to oil.

Policy Responses to Rising Oil Imports

Each of the foregoing risks involves potential economic costs. They include losses to the national economy caused by supply disruptions, recessions brought on by rising oil prices, dollar devaluations necessitated by payments for foreign oil, and the costs of competing with other countries for access to oil. These potential costs constitute the "oil import problem." Solving it will entail minimizing the costs created by oil imports.

The thrust of energy policy since 1974 has been to reduce the costs of imports by reducing imports themselves. Yet the real magnitude of this task is only now becoming apparent. Achieving the administration's suggested 1990 import goal of 4 million barrels per day would require a reduction of 7.3 million barrels per day in oil demand, approximately as much as current imports. A reduction of this size would entail a massive conversion to new sources of energy and a recycling of the nation's capital devoted to energy, entailing large, and often uncertain, economic and environmental costs.

The costs of the policy will be large because imported oil is now cheaper than its potential alternatives, and may continue to be so if the costs of those alternatives rise. The extra costs of the alternatives will be paid either by consumers, in higher prices, or through governmental subsidies requiring higher deficits or higher taxes. The premium may be worth paying, however, since oil imports pose risks. The costs of those risks may not be apparent to consumers, since the economy as a whole pays for them. Disruptions in supply, or a weakening of the dollar, or the military defense of the Persian Gulf are not immediately visible at the gasoline pump or in household fuel bills. It may, therefore, be sensible to allow some subsidization of alternatives to imported oil.

Yet if the oil import problem is defined as the risks posed by imports, rather than the imports themselves, policy need not be confined to import reduction. An alternative goal might be to address the risks by preparing for disruptions in foreign supplies, or taking measures to offset the effects on the economy of higher oil prices or of a devalued dollar. This type of policy would serve the same purpose as import reduction by protecting the economy and society from the inherent risks of imports.

In addition, some policies might be able to reduce the risks associated with any level of oil imports. This type of policy would be concerned with obtaining imports on the best possible terms, as opposed to reducing them. If such a policy could be made effective, it might be desirable to tolerate a higher level of imports and to pay a smaller premium for alternatives to imports. For example, a barrel of oil received in direct exchange for U.S. goods does not pose a serious risk to the dollar. Similarly, diversifying U.S. sources of foreign oil might reduce the risk of a disruption in supply. Thus, policy options can be grouped into three kinds: policies that reduce oil imports; policies that offset losses associated with oil imports; and policies that reduce the risks inherent in oil imports.

Reducing Oil Imports

Reducing oil imports would reduce the risk they pose to the economy. Policies to this end involve substituting alternative energy sources, such as solar or renewable energy forms or synthetic fuel, or reducing demand through conservation or by means of restrictive quotas or fees. With the exception of restrictive policies that enforce import reductions, these policies would provide only long-term relief and would not have significant effects until the end of the decade. Moreover, reducing imports would not eliminate many of the risks they pose. Even at lower levels, imports might still be vulnerable to supply interruptions.

The bulk of the oil used in the United States is employed either in transportation or in space heating. Alternatives to oil can perform these functions. Among these alternatives are synthetic fuels made from coal, grain, or shale.

The latest series of OPEC price increases has given more urgency to the production of synthetics, but significant problems remain. The most promising synthetic is methanol, based on a technology that is generally understood and has been in use, on a small scale, for several decades. Methanol can be used as a liquid motor fuel. It is relatively easy to transport, burns cleanly, is compatible with gasoline, and costs $30 to $40 per barrel. This cost will increase, however, as fuel prices increase with the price of oil. In addition, higher interest rates will increase the financing costs of capital-intensive synfuel plants. Investment in commercial-scale methanol plants, however, has been inhibited by uncertainty over future prices and the initial costs of plant construction. Even if these obstacles were overcome, the long lead times needed to construct a commercial-scale methanol plant would hold 1990 production to less than 1 million barrels per day.

Shale oil would cost $30 per barrel or more before refining. It poses significant environmental problems, including the disposal of spent shale and the leaching of toxic substances into groundwater. The centralized location of shale resources in the West poses logistical problems and threatens to concentrate its environmental consequences in one region.

Grain-based alcohol is a doubtful economic alternative, and its potential for large-scale production is clearly limited.

All of these synthetic liquids have in their favor that they would provide the type of light liquid motor fuel that will be in shortest supply in the future, as the quality of natural petroleum decreases.[12]

Solar heating and cooling carry a resource cost of approximately $40 per barrel in their typical application—no higher than some synthetic fuel production costs. Unlike the synthetic liquids, however, solar applications minimize environmental damage and employ a more advanced technology. Yet solar energy could not replace large amounts of oil quickly because it would first become economic in the South and Southwest, where coal, nuclear-based electricity, and natural gas are used for heating. The use of oil for heating and generating electricity is centered in the Northeast, where it would have to become much more expensive before solar energy would be an economic alternative.

Reducing Oil Demand

Since the 1973-74 OPEC price increases, the United States has made moderate progress in conserving oil; the extent of this progress differs significantly among various sectors of the economy. Industrial energy use has shown a marked decrease over the period, commercial conservation has been less pronounced, and households have shown little response. Yet household uses of oil—including autos, home heating, and some oil-fired electricity—comprise more than half of the demand for oil.

Incentives to encourage households to conserve energy are limited under current law. Few households anticipate owning their homes long enough to justify conservation investments. In addition, household improvements that conserve heating oil and gas produce somewhat intangible returns through lower fuel bills rather than immediate cash rewards. In order to achieve large aggregate energy savings, many millions of households would not only have to recognize the long-term cost advantage of conservation improvements, but would also have to be willing and able to make the front-end cash outlays to pay for them.

The price system has so far not offered much incentive to conserve gasoline. Most of the conservation that has been experienced in gasoline consumption has been achieved through the standards set by the Energy Production and Conservation Act (EPCA), which mandated more fuel-efficient autos, rather than through market incentives. By 1990, EPCA standards will have resulted in as much gasoline savings as would a direct $1.00 per gallon gasoline tax in 1980 that rose at the rate of inflation.[13] Gasoline prices have only now risen to the level at which the costs of fuel-saving innovations are outweighed by the dollar value of fuel savings.

Conservation investments in the commercial and industrial sectors also suffer, as do households, from the somewhat intangible form in which their return occurs. In order to promote additional conservation, new incentives will be necessary. The subsidy inherent in higher incentives would make the cost of conservation higher than the price of the imported oil it saves. However, if the Congress sees fit to subsidize the production of alternative fuels by setting artificially high prices for them, a similar price for conservation may be in order.

Enforced Reductions

Policies that would reduce oil demand through substitution of alternative fuels or conservation require lengthy lead times for investments. Synthetic fuel production, solar and other renewable energy forms, and retrofits of existing energy-using equipment all have the potential to reduce oil imports by a significant amount, but these savings would probably not be realized fully until the 1990s. If the risks associated with the existing level of oil imports are perceived as being so severe as to warrant immediate reduction, then it might be appropriate to restrict imports either through a quota or through imposition of special taxes, despite the possible economic costs.[14]

Reduction through a quota. A quota could be implemented in three ways. The first would be to allow every U.S. importer a prorated share of imports. This, however, would revive many of the inefficiencies found in the former allocation system for domestic oil. Like all other quota schemes, it might also divert supplies to other oil-importing nations, possibly even lowering spot prices and improving availability outside the United States.

A second scheme would be to "auction" import licenses; the number of licenses so auctioned would be limited by the size of the quota. The cost of the import licenses would be passed through to refiners and, ultimately, to consumers. The resulting price increases would be reflected in all domestic oil as well, assuming decontrol. The level of price increases would be equal to an "embargo" on imported oil of similar magnitude.

A third way of implementing a quota would be based on the Adelman Plan, originally put forward by Morris Adelman. The U.S. government would announce how much oil it was going to import and would solicit sealed bids from oil-producing countries to fill the quota. Presumably, this would require the producing countries to compete with each other, thus forcing import prices down.

This last plan, like others that have been put forward to "break up" OPEC, seems attractive in theory, but there is little certainty that it would work. Although OPEC has not yet instituted any formal production-sharing arrangements (even in the glut market of 1977 members used price discounting to equate supply and demand rather than the traditional cartel system), OPEC might decide to prorate, or otherwise divide, production for the U.S. market if faced with a quota. In that event, OPEC would probably reduce production until it was once again equal to demand, as restricted by quotas, so that prices would still rise. This would not only result in higher oil prices but could lead to a greater degree of OPEC control over the world market than it already has. Even if the imposition of quotas forced OPEC to lower its prices on world markets, it could still mean higher delivered prices to the U.S. consumer unless price controls were imposed because of the scarcity quotas would create in the United States.

Reduction through taxes. Demand could also be reduced by taxing imported oil, all oil, or specific products such as gasoline. A tax would induce some reduction in oil demand, but not without cost. Any tax would have both inflationary and recessionary effects, similar to those that are felt when OPEC raises prices by the same amount. This could be mitigated, however, by rebating the resulting revenues through lower personal income taxes, Social Security taxes, or the like.

More limited taxes than those on all oil pose additional difficulties. A tax on foreign oil only would create a windfall for domestic producers (since domestic refiners would be willing to pay domestic producers the market price plus the value of the tax) and might penalize refiners relying heavily on foreign crude. A tax on gasoline only would induce consumers to convert to diesel fuels, which compete with home heating uses for the supply of middle distillates and also create additional environmental degradation.

Demand reduction policy, in sum, may be limited in its effectiveness and expensive to the national economy. For example, this analysis assumes a world oil price of $51.56 per barrel in 1985, with projected imports of 10.1 million barrels per day. Reducing imports to 8.5 million barrels per day in that year would require raising prices to approximately $74.00 per barrel. An import fee of approximately $22.00 per barrel would be necessary, imposing a cost on the economy of approximately $144 billion in 1985 dollars before recirculation. A gasoline tax of $0.50 per gallon in 1979 dollars, rising at the rate of inflation, would reduce demand by 450,000 barrels per day in 1985. Achieving the target of 8.5 million barrels per day in that year would require a gasoline tax of approximately $2.50 per gallon in 1985 dollars.

Offsetting Losses Created by Disruption of Oil Imports

It might not be possible, short of draconian measures, to reduce imports to levels at which their risks become tolerable. Disruptions of supply, for example, can occur at any level of imports. One alternative kind of energy policy would seek to reduce the losses posed by oil imports rather than reducing the imports themselves. A description of these alternative options follows.

The Strategic Petroleum Reserve

One of the most important options is the Strategic Petroleum Reserve (SPR). In order to test the usefulness of the SPR, supply interruptions were assumed with the reserve at various levels. Using the simulation described previously in Chapter 5, if there were a one-year shortfall of 3.5 million barrels per day in 1984, a reserve of 250 million barrels could avert a loss in GNP of between $54 billion and $93 billion. Similarly, a 500 million barrel reserve could avert a loss between $110 billion and $187 billion. The average benefit per barrel of oil in terms of averted GNP loss, net over costs, would be between $216 and $374. These figures do not include either the value of the reserve as a deterrent to politically motivated disruptions, or the revenues derived from selling the oil. The effectiveness of the SPR in averting economic losses would depend on the degree to which the United States could capture all of the benefits provided by the oil, and whether price controls and allocation policies were adopted.

These results indicate that the Strategic Petroleum Reserve would be an effective policy option for decreasing the output losses likely to occur as the result of an oil production cutback. Nevertheless, final assessment of the effectiveness of this policy option must be balanced both against its cost and against the probability of a cutback. The actual value of the Strategic Petroleum Reserve to the federal government can be estimated only through a form of risk analysis, and the probability of a cutback cannot be determined.

The drawback of the SPR is that it requires a large up-front investment in procurement and storage costs—the latter approaching $4 per barrel. One solution might lie in financing the SPR through a public bond issue, allowing private investors to hold title to oil in the reserve as a transferable asset. Whatever the sources of financing, the SPR must be considered an effective option.

Rationing

Another policy that would address the effects of disruptions is rationing. Rationing plans are usually limited to motor fuels, since rationing of heating oil would be difficult to administer and could cause unnecessary hardship. Households consume roughly 5 million barrels per day of gasoline; businesses and government consume 2.5 million barrels per day of gasoline; and the economy as a whole consumes about 1.5 million barrels per day of diesel fuel, primarily in the commercial sector. Since rationing of commercial fuel is thought to have a greater detrimental effect on economic activity than rationing of household uses, it could be focused on the latter. In that case, rationing could allocate shortages of up to 1 million barrels per day. Beyond this level, it is doubtful that rationing would protect the economy from losses.

National Economic Policy

Losses might also be mitigated by macroeconomic measures. Expanding the money supply when foreign oil prices rise might reduce the recessionary effects of such increases, but this would have an adverse effect on the price level in the long run. Similarly, measures like the November 1, 1978, support package might be effective in maintaining the dollar's value in the short run. On the other hand, price controls and allocation programs would not mitigate the direct effects on GNP of oil price increases. In the example cited previously in Chapter 5, an interruption of 3.5 million barrels per day in foreign oil supplies would result in an increase in the general price level of 20 percentage points in that year over the anticipated rate of inflation. Price controls and accompanying allocation programs, in this example, would reduce the inflation increment by only 3.5 percentage points; they would also contribute to the dampening effects of the interruption on the economy.

Reducing the Risks Associated with Oil Imports

The third type of policy would seek to reduce the risks associated with any level of oil imports. For example, a barrel of oil received in exchange for U.S. goods and services would be preferable to one bought with dollars that might not return to the United States. Similarly, policies that would work to expand the supply of oil in countries outside of OPEC would serve the same purpose as those that reduce oil imports from OPEC itself.

A previous CBO publication described the "strategy for oil proliferation."[15] That strategy consists of creating new sources of funds for oil and gas exploration and development in non-OPEC third world countries that may have such resources. Some countries and international organizations have pursued the equivalent of this strategy by setting up "lending windows" or using their national oil companies to stimulate exploration in less-developed countries. The United States could do likewise by extending credits for oil and gas exploration to non-OPEC third world nations and accepting "participation" oil in repayment or joining in international agreements to do so.

Both the costs and benefits of a proliferation strategy are uncertain. It is conceivable that the fund would become self-financing after several years, as initial loans were repaid. Given the possibility of higher real prices for crude oil, repayment would be easier to achieve over time. This program would suffer from the riskiness inherent in all oil and gas exploration, but there is a good chance that it would be cost-free from a budgetary standpoint.

To the extent that a proliferation policy succeeded, non-OPEC oil supplies would expand. This would ease price pressures in the long term and decrease the likelihood of disruption in the international market. Moreover, a governmental lending entity could accept crude oil in repayment, avoiding the outflow of dollars associated with conventional imports. (It should be noted that if other industrialized countries adopted such a policy first, the United States might find its access to crude oil and petrodollar recycling worsened, and be forced to adopt such a strategy as a response.) Finally, exploration aid might be an effective tool in promoting better relations with strategically important countries such as the Sudan, Pakistan, and Egypt.

One class of hydrocarbons that remains underdeveloped throughout the world is heavy oil—oil so viscous that it defies pumping at conventional temperatures and pressures. The world's heavy oil reserves are centered in Venezuela and Canada. The Alberta reserves are now being developed by Exxon, which expects production of about 150,000 barrels per day by the mid-1990s. The Venezuelan reserves, centered around the Orinoco River and Lake Maracaibo, are larger than the Canadian but not yet extensively developed. U.S. energy companies could assist in building such an industry in Venezuela, in exchange for price and marketing guarantees to Venezuela for allowing development. This arrangement could improve the stability and security of U.S. oil supplies.

Another policy would be aimed at existing producers with the ability to expand capacity. Some "banker nations," such as Saudi Arabia, Kuwait, or the United Arab Emirates, have potential production levels above those necessary to finance their national development plans. They have no incentive to expand production beyond that level in return for foreign assets that do not appreciate as fast as the value of their untapped oil. Proposals have been made to create an "indexed" asset that would be linked to the price level or adjusted for dollar devaluation. One variant of this proposal was put forward by former Venezuelan President Carlos Andres Perez, who suggested using surplus petrodollars, matched by funds from the industrialized countries, as the basis of a "Marshall Plan" for all underdeveloped countries. This proposal could not only help to expand the supply of oil from producing nations, but could also improve relations between oil consumers and oil producers. Oil producers would recognize the creation of an indexed asset or a Marshall Plan fund as evidence that oil consumers understand their problems. This might be particularly effective in maintaining the special relationship between the United States and Saudi Arabia.

A different group of policies would aim at improving the recycling of petrodollars and reducing the risk of continued dollar devaluation. One option would be to seek more oil under bilateral agreements, as most other industrialized countries do. This would involve the U.S. government in one of several new roles. The government could repay oil debts through lines of credit established against U.S. goods and services. Alternatively, it could "syndicate" export packages to oil producers in repayment for oil imports that, once again, the government would pass on to U.S. refiners and distributors. Petrodollars could also be recycled through stepped-up capital placement in the United States by OPEC members. This could be achieved by encouraging direct investment by those countries in the U.S. firms and manufacturing assets.

In theory, an oil-for-goods exchange could benefit both parties. Japan has a number of such agreements, covering imports of almost 1 million barrels per day. Canada, England, France, Japan, and Sweden have either made such agreements with Mexico are are negotiating them. Failure by the United States to do likewise could be to its disadvantage if other oil-importing countries secure preferential access to large amounts of crude oil through this method.

Bilateral exchange of oil for commodities would help to stimulate domestic exports, resulting in growth and higher employment. Most other industrialized countries have government entities that promote exports through financing, assembling of export packages, reciprocal marketing guarantees, and other means. While the United States does have the Export-Import Bank, which lends money to other countries to help them import U.S. goods and services, it could do more in this regard. Other industrialized nations also promote exports through regional arrangements with African and Latin American countries. The less-developed countries are rapidly becoming the largest market for U.S. exports because of their high rates of growth. Their demand could be strengthened through increased U.S. participation in international development institutions, increased U.S. support for the Agency for International Development, or through the Export-Import Bank.

The policy has possible drawbacks. First, the terms of exchange between oil and its reciprocal commodities must be fixed over time, or some mechanism must be created to accommodate changes in their relative prices. Second, in the event of a major disruption in oil supply, bilateral contracts might not be honored.

Many believe that a formal dialogue between the oil-producing and oil-consuming countries is inevitable. A formal French-Kuwaiti proposal exists for such a conference between OPEC countries and the members of the European Economic Community. Such a conference could go beyond the questions of oil price and availability to issues such as whether to abandon the dollar as the denominator of oil prices, questions of trade between the industrialized and less-developed nations, and political solutions to the Palestinian question. Dialogue between oil producers and consumers could also take a regional form. Several proposals have been made for a North American "common market" that would accelerate trade between the United States and its neighboring energy producers, Canada and Mexico. This is a particularly attractive possibility with regard to Mexico, for which trade concessions have been a political priority.

Many regard this type of dialogue as as inevitable precondition to stability in the oil market. As Alberto Quros, president of Venezuela's state oil company, Maraven, has stated:

> Interdependence is the only way out, but only the developed nations can really take the steps leading to partnership with the developing world. . . . What is needed is a long-term understanding under which OPEC would now supply the volumes the world needs at prices that consumers can afford, in exchange for future considerations, financial assets, market prices, technology transfer and trade, in which the OPEC producers can have some confidence.

NOTES

1. Natural gas liquid is natural gas that occurs in liquid form and is produced in conjunction with natural gas. It is incorporated into the refining process along with crude oil.

2. Oil in the ground, or oil in place, includes the total known amount, including that which cannot be recovered economically by today's technology.

3. For more detail on the concept of high- and low-absorbing countries, see Dankwart A. Rustow, "Political Factors Affecting the Price and Availability of Oil in the 1980s" (New York: Petroleum Industry Research Foundation, 1978).

4. See Congressional Budget Office, The Economic Impact of

Oil Import Reductions (December 1978) and Strategic Petroleum Reserve: An Analysis (1980). Washington, D.C.: Government Printing Office.

5. The results of the analysis are presented in annual terms because it is difficult to estimate precisely the impact of a shorter supply interruption. There is too much uncertainty as to how conservation measures, fuel switching, the drawdown of the Strategic Petroleum Reserve, and the level and drawdown of oil pipeline inventories would combine to mitigate the effect of the production limitation. For an oil production limitation of less than a year, the impact on real output (as well as prices and unemployment) is assumed to be linearly related to the annual results. For example, a $22.5 billion loss to the economy would result from a six-month supply interruption, compared to a $45 billion loss during a yearlong cutback.

6. The price controls assumed in this analysis permit price increases sufficient to compensate for increased costs of production resulting from bottlenecks and inefficiencies in producing goods and services. The inefficiencies are obviously greater at higher levels of oil shortfall, leading in turn to more rapid increases in prices. Finally, it should be noted that a fuller discussion of the interaction between oil prices and GNP can be found in a number of CBO publications. See for example, President Carter's Energy Proposals: A Perspective (June 1977), chap. 2; and Recovery: How Fast and How Far? (September 1975), chap. 5, Washington, D.C.: Government Printing Office.

7. For more details on U.S. economic interdependence with other countries see Klaus Knorr and Frank W. Trager, eds., Economic Issues and National Security (Lawrence, Kansas: Regents Press of Kansas, 1977).

8. For more background, see Aaron Wildavsky, ed., Energy and World Politics (New York: Macmillan, 1975); and Edward N. Krapels, "Oil and Security Problems and Prospects of Importing Countries," Adelphi Papers no. 136 (London: The International Institute for Strategic Studies). See also Szyliowicz and O'Neill, The Energy Crisis and U.S. Foreign Policy (New York: Praeger, 1975); Yager and Steinberg, Energy and U.S. Foreign Policy (Cambridge, Mass.: Ballinger, 1974).

9. The leading exposition of this school of thought can be found in Edward Friedland, Paul Seabury, and Aaron Wildavsky, "Oil and the Decline of Western Power," Political Science Quarterly 90 (Fall 1975). The same authors have also published The Great Detente Disaster: Oil and the Decline of American Foreign Policy (New York: Basic Books, 1975).

10. It should be noted that U.S. analysts believe that the Soviet bloc will become a net importer of oil in the 1980s. CBO believes

that the Eastern European countries will also become net importers in the 1980s. See Central Intelligence Agency, Prospects for Soviet Oil Production (April 1977). The CBO estimates are discussed in Chapter IV of this report. Issued by the Central Intelligence Agency in Series ER 77-10270, Government Printing Office, Washington, D.C.

11. Knorr and Trager, Economic Issues and National Security.

12. This is true worldwide. Superior, lighter oil has been over-produced in proportion to its occurrence. Many producers are now trying to produce heavy and light oil in proportion to reserves. This amounts to a trend to decreasing crude oil "quality," as measured by the ability to refine oil into light products such as gasoline.

13. See statement of Alice M. Rivlin in Automobile Fuel Economy, U.S. Senate Committee on Commerce, Science, and Transportation, Hearings before the Sub-Committee on Science, Technology, and Space, United States Senate, 95th Congress, 1st Session, July 12 and 14, 1977. Serial Number 95-38, Washington, D.C.: Government Printing Office; and Congressional Budget Office, Preliminary Projections of Fuel Savings and Revenues Associated with Increased Taxes on Motor Fuels (1979). Washington, D.C.: Government Printing Office.

14. This discussion of import fees and quotas is based on Congressional Budget Office, Direct Federal Action on Oil Imports (1978). Washington, D.C.: Government Printing Office.

15. Congressional Budget Office, A Strategy for Oil Proliferation: Expediting Petroleum Exploration in Non-OPEC Developing Countries (1979). Washington, D.C.: Government Printing Office.

II

Estimating Oil and Gas Resources

INTRODUCTION

The hazards of predicting available energy resources and energy consumption patterns were stated clearly in Chapter 1. Indeed, the complexity of the process and the associated conflicting estimates and projections undoubtedly have contributed to the widespread popular misconceptions over the existence, nature, and extent of energy problems. In addition, the uncertainty concerning energy resources certainly complicates the efforts by policymakers to develop a consistent and effective energy policy.

The two chapters in this part further illuminate the complexities of the problems of estimating resources. In Chapter 2, John J. Schranz, Jr., suggests that most past discussions of oil and gas resources have been exercises in confusion and uncertainty. It is little wonder the public and elected officials are confused and frustrated when experts disagree on critical factors such as terminology and the basic methodology to use in the estimation process. When these problems are coupled with those related to forecasting production rates for two or three decades and predicting the impacts of economic and technological changes during the same time span, the problems become even more evident to the reader.

The difficult task of looking into the future becomes even more complex when we attempt to predict the human behavior and decision makers and the future actions of governments of nations throughout the world. The rapid changes and upheavals that have occurred in the international political arena just during the past decade stand as a stark reminder of the instability that characterizes many nations throughout the world. This factor of instability and the problems related to recurrent natural disasters and unpredictable acts of violence should serve as a cautionary reminder to all of the precarious nature of estimating energy resources. As world oil production nears or reaches its peak in the decade of the 1980s, the increased competition for scarce energy resources will severely test the abilities of nations, alliances, and policymakers. As Stansfield Turner points out in Chapter 3, a critical task of the leaders of all nations will be to prevent the competition from becoming mutually destructive. As such, the chapter concludes with an overview of the economic, political, and military implications of the world energy situation in the 1980s.

2

OIL AND GAS RESOURCES—
WELCOME TO UNCERTAINTY

John J. Schranz, Jr.

Until 1973, the American public was accustomed to glad tidings
about U.S. oil and gas resources. If you read the business sections
of newspapers or followed the trade and professional publications,
you were aware that the forecasts became increasingly optimistic
over the years. There were some less sanguine estimates from those
who looked at the ever-declining curves of oil field production and
projected rising costs through time. But these more cautious projections appeared to be overshadowed by the upward path of U.S. oil
production.

The undercurrent of concern during the 1960s over declining
exploratory activity in the United States elicited little real attention
outside of the oil and gas industry itself and a small circle of petroleum specialists. It was easy for others to treat these worries as
merely the customary favorable treatment by Congress on taxation,
incentives, or protection from foreign competition. However, the
major disturbance caused by the Organization of Petroleum Exporting
Countries (OPEC) oil embargo in 1973 brought an immediate end to
this lack of public attention.

In 1975, a report by the Committee on Resources and the Environment (COMRATE) of the National Academy of Sciences, based on
a review of contemporary estimates, stated that of the original stock
of crude oil and natural gas liquids (249 billion barrels) only 113

The editors wish to thank Resources for the Future for permis-
sion to reprint this special issue from Resources, no. 58, Resources
for the Future, March 1978.

TABLE 2.1

Changing Perspectives of U.S. Oil and Gas Resources

Forecast in Year	Original supply of recoverable reserves	
	Oil (billion barrels)	Natural Gas (trillion cubic feet)
1948	110	—*
1952	—	400
1956	300	856
1965	400	2,000
1969	—	1,859
1970	432	—
1972	458	1,980
1975	249	1,227

*Data not available.

Note: Unfortunately, any sampling of estimates encounters variations in the treatment of past production, recoverability, and the inclusion of natural gas liquids. These have been chosen, or adjusted when possible, to make the totals roughly comparable regardless of year of estimate.

Source: Compiled by author.

remained to be discovered. For natural gas 530 trillion cubic feet (of an original 1,227 trillion) remained. (See Table 2.1.) This marked the end of general optimism both in industry and government about the future U.S. oil and gas resource position. To have this unwelcome news appear in the midst of the oil and gas industries' postembargo clamoring for high prices resulted in both public confusion and distrust. With respect to natural gas, the winter crisis of 1976-77 caused renewed doubts and confusion among the public, the media, and members of government.

A MATTER OF DEFINITION AND CLASSIFICATION

An oil or gas reservoir is not a subterranean cavern filled with oil and gas that we empty like a huge storage tank. During geological time, various mixtures of crude oil, natural gas, and salt water were formed and moved about in the interconnected minute pores of certain kinds of rock where they have remained trapped. When the driller's bit penetrates the rock, natural pressures cause a slow migration of the fluids toward the well bore. The well operator may decide initially, or eventually, to give the flow of oil and gas an assist through the application of the sucking action of a pump, or by fracturing the rock around the well, or by injecting water, chemicals, heat, or gases into the rock. To understand this production process, three things must be kept in mind: first, the flow of fluids through rock pores is a function of the physical forces at work; second, the quantity resulting from additional effort gradually diminishes, just as wringing a wet rag produces less and less water; and third, the only actual measurement that can be made is of the oil and gas produced at the surface— all other information about the reservoir is estimated.

Once the physical characteristics of an oil and gas resource system are appreciated, the complexity of the question "How much oil and gas do we have?" becomes more apparent. Any response can be no more than a judgmental estimate. Intelligent communication about oil and gas resources becomes exceedingly difficult unless both the questioner and respondent understand what kind of data they are using.

The total oil and gas resource base includes all unproduced natural oil and gas hydrocarbons that may exist. It is worth repeating that past production, that is, the quantity already delivered by this system, is our only actual measurement. The quantity of oil and gas is gone forever. References to original oil and gas in place mean the sum of both the remaining oil and gas plus all that has ever been produced.

The productivity capacity of the United States is the amount of oil and gas that can be produced from existing wells during a specified period of time under specified operating conditions. The totality of physical oil and gas in the earth but not yet produced from the continental crust to a depth of perhaps 50,000 feet is sometimes called the oil and gas resource base. There are four kinds of oil and gas found in this resource base. The first kind consists of oil and gas that has already been found and is considered producible under present prices using current technology. These quantities are customarily known as reserves. The immediately producible portion of these reserves—the oil and gas that will flow from wells in developed reservoirs, the quantity of which can be estimated with considerable accuracy—is classified as proved reserves. The balance, or unproved reserves, has been discovered but cannot be estimated with as great accuracy and may require additional drilling and development.

Oil and gas that has been discovered but in the judgment of the operators cannot be produced under current prices with existing technology are known as subeconomic resources. There are two kinds of subeconomic resources. First, the unrecoverable, high-cost portion of oil and gas currently left behind in producing reservoirs. Second, oil and gas in other reservoirs that have been found but are not now producing or have been abandoned because they would cost too much to produce due to size and other problems.

What remains is the oil and gas that requires further exploration. Exploratory drilling has not proceeded to a point where there is physical evidence of the actual presence of this oil and gas. There is only expectation, and estimates of undiscovered oil and gas are based solely upon geologic and engineering extrapolation. This requires the use of geological and geophysical data rather than using physical data based upon the actual existence of the oil and gas. It is possible to subdivide undiscovered resources into economic and subeconomic quantities, but to do so requires the analyst to make some sort of assumption about prices and technology conditions. Present prices and technology are frequently assumed despite the fact that the oil and gas, when actually discovered, will be produced under future conditions of price and technology.

The final portion is other occurrences. This includes any oil and gas left behind that is not expected under any future circumstances to be worth the effort or cost of production, as well as deposits that are considered too small to either find or produce if found. Finally, this category is a convenient place to account for other forms of oil and gas hydrocarbons about which either little is known or production technology is so immature that economic and technologic judgments cannot be made, even though large quantities may be involved.

ESTIMATION OF RESERVES

As the drill bit penetrates a rock reservoir for the first time and finds oil and gas, the first questions asked are how much has been found and can it be produced economically. The initial well provides limited information about the rock strata that have been penetrated and nothing about strata that are below the bottom of the well. Once a layer of rock containing oil and gas has been found, the approximate thickness of the bed at that point—anything from a few to hundreds of feet—is known. A core of rock is usually taken from the bed. Electrical and other measurements are taken inside the hole. All of these data provide information about the porosity and permeability of the rock and the amounts and kinds of fluids it contains. The initial flow of a new well provides information about the production rate, pressure, and other physical data.

At this point, a preliminary judgment on how much oil and gas have been found can be made based on the flow from a single well, a rock sample a few inches in diameter of a multiacre reservoir, a map of the surface geology, and a seismic "shadow picture" of the structure holding the oil and gas thousands of feet below the surface. Obviously, the first estimate cannot be very precise. Yet based on this one well and past experience with the kind of reservoir that appears to have been found, the engineer makes a judgment. This estimate may range from the least amount of oil and gas that appears to have been found to the outer limit of what the reservoir might ultimately produce if the buried structure is entirely filled with oil and gas. The scientific guesswork about a reservoir hundreds of acres in size is useful but extremely crude.

Even before a well is drilled, companies will appraise the potential of a new region to help them determine whether or not exploratory wells are worth drilling. Once a well has been successful in finding oil and gas, two new estimates can be made: first, an estimate of the minimum amount (the proved reserve) that seems to be producible by that well and, second, a less certain estimate of what might be the ultimate potential of the entire field. As more wells are drilled and additional production data are gathered, the proved reserve estimate may be revised up or down. The expectation of ultimate production can also change upward or downward, usually over a much wider range than that of the proved reserve figure—several multiples are common. For a typical field it takes approximately five or six years before the proved reserve estimate of remaining oil plus past production begins to approach a true estimate of ultimate production. In other words, it takes a number of years before there emerges a reasonably accurate estimate of what has been discovered in toto in a reservoir. The exact amount of producible oil or gas is not known until the field is permanently abandoned and that oil or gas has been measured as past production.

In addition to a company's estimates of proved and ultimate, other appraisals may be made by a producing company during the life of a field for various purposes. Estimates based on well logs and other data are commonly used by banks for making loans. Information is also released to the press about the importance of new discoveries. The Securities and Exchange Commission expects that companies will release information to stockholders about their holdings and expectations. And, finally, the many kinds of information required by government agencies lead to a number of estimates being provided by a variety of federal offices. Considering the array of purposes for which reserve estimates are made and the constant revision of most of these through time, it is not surprising that various reserve reports may appear to be in conflict.

The proved reserves of oil and gas represent only a small portion of the total oil and gas resource base that remains unrecovered in the United States. Yet, these proved reserve data sometimes receive more attention, and in recent years have prompted more controversy, than the more significant resource estimates of undiscovered oil and gas. For crude oil, proved reserves represent the stock of immediately producible oil from existing wells. The oil producer knows that the amount of oil or gas that can flow in a given year from producing wells is physically linked to the number of wells available and the quantity of oil and gas still remaining in the reservoir. Thus, proved reserves for many years have been the industry's empirical indicator of current capability, not a measure of the total supply of oil and gas left for the future. Equating proved reserves with "years of supply" is particularly misleading.

Unfortunately, there are no regular government or industrywide efforts to report on what is known as indicated or inferred oil and gas. The American Petroleum Institute (API) does report on additional reserves of oil that could be produced from secondary recovery projects but that are not yet fully evaluated at the time of the proved reserve estimations. An industry-sponsored effort called the Potential Gas Committee has included this portion of the gas reserves in its occasional reports on gas resources. The Federal Energy Administration in its 1975 survey of operators had hoped to go beyond merely proved reserves, but its final report only included oil from secondary and tertiary projects and not the less certain oil and gas quantities. Currently, the new Department of Energy is again considering how to define and request data on oil and gas reserves that are not reported as proved.

The U.S. Geological Survey in its 1975 Circular 725 relied upon the use of a statistical ratio devised by M. King Hubbert to account for indicated reserves. This ratio is based on the historical relationship of the amount of oil and gas that has been added through extensions and revisions to proved reserves during the typical life of reservoirs. The relationship shows that approximately 80 percent more oil and gas will be produced from known fields than is currently being reported as proved. Although proved reserves data are considered to be accurate within plus or minus 20 percent, this refers to the oil and gas expected to flow from existing wells. On the average, almost twice as much oil and gas will ultimately be found in these fields once their true size or limits become fully identified. This is not an intentional understating or "hiding" of reserves, but merely a reflection that the definition of "proved" limits the estimators to the drilled portion of the field.

One must be aware that when estimates go beyond proved, accuracy deteriorates rapidly, with errors of perhaps 50 percent or more for indicated reserves (mostly oil and gas resulting from further

development of the reservoir) and amounting to perhaps several multiples for inferred reserves (oil and gas resulting from the discovery
of additional reservoirs within the same field).

To obtain more information on what lies beyond proved reserves
in known fields involves a reservoir-by-reservoir examination of considerable magnitude. Considering the range of judgment involved and
the unavoidable approximations, the added information obtained may
not be worth the cost of acquiring it. In addition, there are problems
of handling proprietary data, which in any event, would be diverse,
constantly changing, and of unknown quality and usefulness.

In the final analysis, to know that indicated and inferred U.S.
reserves are considered to be 2.4 times our proved reserves instead
of 1.8 times has little significance in determining our policies with
respect to oil and gas. The more important questions are found in
the categories of subeconomic and undiscovered oil and gas resources.
It is upon these quantities that our energy future depends most heavily
in the medium term.

ESTIMATION OF SUBECONOMIC RESOURCES

Subeconomic resources include all of the oil and gas in known
reservoirs that is not producible by present technology at present
prices, but may become producible in the future with improved technology or higher prices. It should be noted that a simple downward
movement in prices or other incentives can cause some reserves to
be reclassified, at least temporarily, as subeconomic resources.

The investment decision in oil production is a balancing of the
total quantity to be recovered, the various costs, the rate at which
the oil will be recovered, and the price at which it can be sold. Once
the decision is made on how many wells to drill and what recovery
technology to use, that is, natural flow, pumping, water flooding,
injection of steam, or other methods, the amount of oil that will be
recovered and the rate at which it will reach the surface are limited
within a fairly narrow range. To change that plan, additional investments must be made in drilling additional wells or in altering the
production methodology being used. Such a change in the production
scheme will be adopted only if the faster production or greater recovery can justify the extra cost.

Thus, an increase in the price of oil or gas may not be adequate
to change the plan for the operation of a field already being developed.
The only effect of higher prices in that event may be to permit the
reservoir to decline to a lower level of daily output per well before it
is abandoned because of low oil flow or gas pressure. This additional
quantity of oil and gas produced in the later life of a reservoir may

only involve a 1 or 2 percent increase in the ultimate recovery because most of the oil or gas left behind is entrapped in the reservoir and could only be recovered by the use of a different technology. However, higher prices that occur before a production plan is fully implemented in a new field can lead not only to later abandonment but to higher recoveries because the prices can be reflected in a timely investment in a modified development scheme.

The appearance of significant improvements in production technology or markedly higher prices can justify modifying the way new reservoirs as well as fully developed fields are being operated. Even an abandoned reservoir can be reopened, although this is less likely because of the expense. It should be noted, however, that new methods of enhancing the recovery of oil and gas should not be viewed as applicable to all kinds of reservoirs. How successful a new technology can be employed is determined by the kind of structure and natural energies in the reservoir as well as the kind of rock that is found in it.

There has not been much experience in estimating the size of the national subeconomic resources of oil and gas because in the past the opportunity to discover new and plentiful oil and gas resources has always seemed more attractive to industry. Even for known fields, the estimation of subeconomic resources is complex. First, the estimator must face uncertainty about new and perhaps untested technology. Second, there is need to deal with the effect of price on production using established technology, as well as what price is required to make new technology commercially feasible. Third, there is a lack of information on exactly how much oil or gas is left in the reservoir to be recovered by new technology. Finally, if the data are to have meaning, there is need to deal with the problem of the time over which these prices and technology can be assumed to occur.

It is perhaps surprising that the amount of oil left behind in a reservoir is uncertain. Depending upon the kind of reservoir and the years during which it was exploited, oil recovery from the initial development plan used can vary anywhere from 10 percent to 80 percent of the oil estimated to have been in place originally. In some cases, reservoirs have been reworked with a secondary production technology long after primary methods were begun. More recently, developed fields tend to be exploited by several integrated methods. Since oil in place, recoverable reserves, and a reservoir's recovery factor are all parts of the same equation, it is apparent that estimates for two of them allow the derivation of the third. Thus, if greater production from a reservoir is achieved than originally expected, one is never sure whether the cause is more oil in place, more reserves, or a higher recovery factor.

There is some indication that the overall recovery factor for oil in the United States did not improve very much during the 1960-75

TABLE 2.2

National Petroleum Council Report on Estimates of
U.S. Enhanced Oil Recovery Potential

Source of Estimate and Price Assumption	Potential Recovery (billions of barrels)	Production in 1985 (millions of barrels/day)
National Petroleum Council Report		
(1976 dollars)		
$5	2.2	0.3
$10	7.2	0.4
$15	13.2	0.9
$20	20.5	1.5
$25	24.0	1.7
GURC[1]		
(1974 dollars)		
$10	18–36	1.1
$15	51–76	—*
FEA/PIR[2]		
Business as usual, $11	—	1.8
Accelerated development, $11	—	2.3
EPA[3]		
(1975 dollars)		
$8–12	7	—
$12–16	16	—
FEA/Energy Outlook[4]		
$12	—	0.9
FEA[5]		
(1975 dollars)		
$11.28	15.6–30.5	1–2

*Data not available.

Sources:

1. Gulf Universities Research Consortium Reports, no. 130, November 1973, and no. 148, February 28, 1976.

2. Project Independence Report, Federal Energy Administration, November 1974.

3. The Estimated Recovery Potential of Conventional Source Domestic Crude Oil, Mathematica, Inc., for the U.S. Environmental Protection Agency, May 1975.

4. 1976 National Energy Outlook, Federal Energy Administration.

5. The Potential and Economics of Enhanced Oil Recovery, Lewin & Associates, Inc., for the Federal Energy Administration, April 1976.

period for several reasons. U.S. production shifted from regions with naturally higher recovery potential to areas with poorer recovery potential. Early estimates of the quantity of oil to be recovered by primary recovery techniques were probably overstated or, conversely, the oil in place may have been understated. Finally, there is a tendency to use a standard recovery factor in relating future production expectations to oil in the reservoir. Each of these tendencies could contribute to the assumption that the U.S. recovery factor has remained near 30 percent for many years. Realistically, it must be concluded that the true national recovery of oil is an unknown percentage.

The current interest in enhancing petroleum recovery by injecting heat, CO_2, or chemicals has led to more vigorous examination of subeconomic resources than ever before. Our major oil regions have been examined in terms of the amenability of the various kinds of reservoirs to newer methods for increasing recovery. Although some optimistic suggestions have been made that U.S. recovery could be increased from an assumed 32 percent to ultimately 60 percent, more modest near-term goals are now being set for the upgrading of some of our subeconomic resources to reserves. These suggest an overall increase of perhaps 5 to 8 percentage points in the U.S. recovery factor may be possible.

The uncertainty of how much subeconomic oil and gas may be produced was illustrated in 1973 by a report of the National Petroleum Council (NPC). (See Table 2.2.) The additional oil from enhanced recovery, according to the NPC, could be as little as 7 billion barrels, under a price assumption of $10 per barrel (1976), but this would increase to 24 billion barrels at $25 per barrel. The effect on the rate of annual production would also vary. At the higher price level, U.S. oil production could be 3.5 million barrels per day greater in 1995. The uncertainty in the estimates is reflected in the judgment that the higher 24 billion barrel amount is merely the central value of an estimate ranging from as little as 12 billion or as much as 33 billion barrels. In contrast, another study viewed the outer limit of enhanced recovery at 76 billion barrels at $15 per barrel (1974). Despite the fact that enhanced recovery deals with "discovered" oil in known fields, this does not narrow the range of uncertainty. Technological and economic forecasting of recovery is a source of frustration equal to that of estimating the undiscovered.

ESTIMATION OF UNDISCOVERED RESOURCES

Unlike manufacturing, or some kinds of mining operations, the capacity to produce petroleum is not a constant. To avoid a decline in national production, there must be continuous drilling and development. The process of continual annual replacement of what we have produced is heavily dependent upon the magnitude of our undiscovered oil and gas resources.

The potential size of these resources is usually evaluated in one of three ways. There is the geologic or volumetric approach, which attempts to make a direct estimate of the quantity of oil and gas remaining to be discovered and recovered. No attempt is made to show when or if these resources will be produced. The second approach is that of the engineer-manager who makes projection of the drilling, discovery, and production process. These future production profiles implicitly suggest the amount of recoverable oil and gas that is left in the ground. The third methodology is that of the economist who uses the equations in his model to suggest what future supply can be achieved by the oil and gas producers as they respond to price changes. Like the engineer-manager, the econometrician may indicate the quantity of remaining oil and gas resources in his model implicitly rather than explicitly.

The volumetric approach. The geologist's volumetric estimate is essentially what the name suggests. The total volume of sedimentary rock suspected to contain petroleum and natural gas is calculated for the entire United States, region by region. Based upon past geological knowledge, an estimate is made as to the total oil and gas that may exist in these rock volumes. It is quite apparent that this is a subjective judgment linked to past experience. Underestimates are possible since past experience does not readily account for unknown types of occurrences or future improvement in the ability to detect and produce oil. In contrast, since there is evidence that better areas and larger pools are found first, unexplored regions may prove to be less prolific.

The volumetric determination of the oil and gas that exists in the ground may not be the only calculation. The quantity of oil and gas in place in the rock strata only has economic meaning in terms of the proportion that is both discoverable and producible. The quantity of oil and gas eventually captured depends upon future effort, the effectiveness of the search, the size and depth of the reservoirs, and the economic and technical capacity for producing it.

A careful examination of past geologic estimates reveals that it is rare for the same type of resource concept to be involved. Total oil and gas originally in place, oil and gas remaining in place, discoverable oil and gas in place, undiscovered commercial accumulations of oil and gas, or recoverable oil and gas under given economic and technologic conditions are markedly different quantitative concepts. Unfortunately, the authors of petroleum resource reports all too frequently are obscure about what they have estimated, their methodology,

and their assumptions. Yet all of these numbers are generally identified as estimates of "the oil and gas resources" of the United States.
The unsuspecting recipients of this information must then puzzle over
how one expert can say that the oil resources of the United States are
50 billion barrels and another, with seemingly equal confidence, provides an estimate of 1,000 billion barrels.

If one reduces all of these various estimates as best he can to
a common base, such as the quantity of undiscovered oil that is discoverable and producible at prices as of a certain date with an assumed
30 percent recovery factor, then the wide differences begin to shrink
drastically. An estimate that appeared to be twenty times as large as
another suddenly is only twice as large. Once reduced to a common
base, there remain understandable differences in judgment between
two analysts who possess varying degrees of optimism about what is
still to be discovered. But this kind of comparison is not available to
the casual reader who cannot know that one geologist has estimated
all of the oil in the ground, another has assumed an optimistic 60 percent recovery factor, and another uses the current 30 percent recovery factor.

Geological resource analysis took on a new dimension with the
1975 publication by the U.S. Geological Survey (USGS) of Circular 725,
Geological Estimates of Undiscovered Recoverable Oil and Gas Resources of the United States. This was a major scale-up in the USGS
effort and involved a whole team of geologists working for a number
of months. It entailed not only the use of traditional volumetric information, but incorporated sophisticated statistical integration of subjective judgments about each of 102 oil and gas provinces. The end
product was a probabilistic appraisal of undiscovered, recoverable
oil and gas.

Experimentation with this type of delphic approach has been
going on for a number of years. Companies and various research
groups have searched for a way to combine the various judgments of
experienced individuals into a numerical expression of the probability
of finding oil and gas. Circular 725 was the first attempt by the federal government to try this approach (see Figure 2.1). Single number
estimates that suggested a precision that does not exist have been
abandoned. The public and Congress may now have to become used to
resource estimates that indicated there is a 95 percent chance there
may be a minimal quantity of oil resources but also a 5 percent chance
that there could be quite a bit more. For example, the USGS estimates
that there is a 95 percent probability that the remaining undiscovered
recoverable oil reserves will be at least 50 billion barrels, but only
a 5 percent probability that they will be as large as 127 billion barrels.
Outside of these ranges there still remain low-level possibilities that
a new province may have no oil or gas at all, or that it may contain

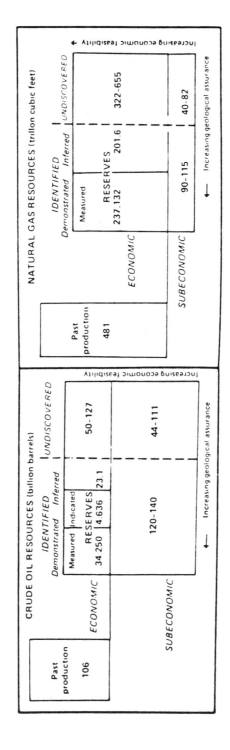

FIGURE 2.1

USGS Estimates of Crude Oil and Natural Gas Resources of the United States, December 31, 1974

Source: Compiled by author.

an undetected Middle East. Only the drill can tell. Some cautious individuals still prefer not to try to attach numbers to what they consider immeasurable quantities.

A ruler placed on the graph of past production and the cost per barrel of any well or field always provides a dismal picture of a downward trend in the absence of new discoveries and technology. In contrast, projections made by individual companies or industry groups showing increasing future production are illustrations of how additional investment in exploration, drilling, or applications of new technology can cause the aggregate production of oil or gas to increase in the future despite the fact that the older wells are declining and future efforts will face greater costs per barrel produced per foot drilled.

Differences in drilling activity can lead to various perceptions of the future. By their very nature, these projections of future production, based on a specified amount of additional effort or investment, are expansive. These speculative futures may or may not incorporate a judgment as to whether the remaining oil and gas resources in the ground are sufficient to provide for these annual flows.

Many government or industry projections of future production are not primarily designed to deal with ultimate size of our oil and gas resources. Nonetheless, they still may foster a public belief that resources are adequate in size to meet the projected goals. In addition, there may be only minimal attention paid to the price required to elicit the necessary investment. And whatever that price may be, the accompanying alterations in the demand for oil and gas, given that price, may not be addressed at all.

ECONOMETRIC MODELS

The third group to deal with the future supply of oil and gas has been the economists. By professional instinct and training, they initially turn to the marketplace as the starting point for their analysis. Their facility for portraying relationships by mathematical equations, combined with the ability of modern computers to provide rapid and complex calculations, has led to the use of econometric models. Normally found within these models are equations that relate the supply of oil and gas to exploratory and development efforts prompted by changes in price.

Econometric models have their own special link to the past. The response, or elasticity, of oil and gas supply to price must be judged in large measure in terms of historical data, despite the realization that in each future year we will deal with a different segment of the original resource. Future resources may very well differ in character and, as a consequence, in cost from those discovered in the past.

TABLE 2.3

Econometric Simulations of Phased Deregulation of Natural Gas

Year	Total Reserves (trillion cubic ft.)	Production (trillion cubic ft.)	Demand (trillion cubic ft.)	New Contract Field Price (cents per thousand cubic ft.)
1972	233.4	23.3	23.5	31.7
1973	227.8	23.6	24.3	34.7
1974	222.9	24.3	26.3	39.7
1975	222.3	26.4	28.7	64.7
1976	226.1	27.6	30.4	71.7
1977	233.9	28.6	31.9	78.8
1978	245.8	30.2	32.9	85.9
1979	258.6	32.1	33.7	93.1
1980	271.2	34.1	34.2	100.3

Source: P. W. MacAvoy and R. S. Pindyck, Price Controls and the Natural Gas Shortage (1975).

Economic behavior patterns of operations conducted on vast federal leases in 1,000 feet of water are not the same as those encountered in the private farmlands of Kansas. Nor will the response of gas supply to a doubling in price (in constant dollars) be the same when it starts at 10 cents per thousand cubic feet (MCF) as when it starts at $1. A reason to question further the future validity of past experience is to recall that much of the past was characterized by smaller movements in the price of oil and gas relative to other prices, and that for the most part this was downward not upward.

CURRENT EFFORTS

In making resource estimates geologists, engineers, and economists are all to some degree projecting past experience into the future. Insofar as the past does not adequately represent the future, their estimates are likely to be in error. In addition, each profession, starting from its own particular analytical framework, is the victim of a certain amount of tunnel vision. The geologist prefers to perform his task in a price-free, time-free fashion. The engineer may ignore

TABLE 2.4

PIES Model Natural Gas Production Reference Scenario

Assumed World Oil Prices $/bbl	1985 Domestic Production (trillion cubic ft.)			1985 Average City Gate Price ($ per thousand cubic ft.)
	Nonassociated Gas	Associated Gas	Total	
$8	16.3	4.1	20.4	$1.79
13	17.4	4.9	22.3	2.03
16	17.4	5.1	22.5	2.07

Note: The reference scenario is a market clearing price at which supply and demand are at equilibrium in an uncontrolled market or deregulated condition.

Source: Federal Energy Administration, National Energy Outlook (Washington, D. C.: Government Printing Office, 1976).

resource constraints and economic reactions in his production model. The econometrician may demonstrate what market price is necessary to reach an equilibrium point but in so doing may violate the time sequence or engineering requirements needed for the process to be accomplished, given the magnitude of the remaining resources and national capabilities.

It is not suggested that the various analysts are totally unaware of the limits of their work. More often than not the problem is the difficulty of trying to link all dimensions of the resource system into one model or into one forecast. Moreover, the purposes being served may not demand a complexity that exceeds available time and financial resources.

The Federal Energy Administration (FEA), in projecting the needs of the nation by 1985 for Project Independence, initially employed the committee approach to the problem; so, too, did the National Petroleum Council. However, subsequent in-house work by the FEA staff on the 1976 National Energy Outlook (NEO) led to the development of the complex computer model (PIES). This effort has been an excellent illustration of the long and difficult task of attempting to introduce the many dimensions of energy into one integrated analysis. The many scenarios developed for the National Energy Outlook required an analysis of demand, supply, finance, the environment, the national economy, and international aspects.

This should not be construed as suggesting that the ultimate model is now available to the new Department of Energy. A close examination of PIES reveals that the tie between energy and the national economy tends to be one directional. In the 1976 report, environmental and international aspects were not introduced as specifically as one might desire. The model reflects the many imperfections in our understanding of the behavior of energy demand in the marketplace. The resource component of the model is still the familiar 1973 data from the USGS Circular 725. Perhaps most important to the user is the fact that the PIES model does not generate a single forecast, but rather as many forecasts are there are policy combinations that an administration wishes to test (see Table 2.4). It is easy to overlook, in the copious statistics and discussions of the model and its scenarios, that much reliance has been placed on a few key sources of data or relationships. Thus, to whatever extent Circular 725 is limited in its perspective of U.S. oil and gas resources, the National Energy Outlook series is equally limited.

Since so many analyses have come to depend upon it, the further work of the USGS has become extremely critical. Currently, it is hoping to refine its presentation of probability data on undiscovered oil and gas resources so that the full range of potential resources within the hypothetical and speculative categories is more apparent. This will allow for an appreciation that beyond the 5 and 95 percent probability boundaries there still remain possibilities for either zero finds or major discoveries for which past experience has not prepared us. Recent interest on the part of the National Petroleum Council and other groups in enhanced recovery will now permit the survey to be somewhat more specific about the magnitude of subeconomic, discovered resources. In addition, the presentation of data on indicated and inferred reserves (reserves beyond proved) in known fields may be expanded by the survey.

OTHER OCCURRENCES

Other occurrences are frequently the source of possible deception about the size of the nation's usable oil and gas supplies. Billions of barrels of oil in low-quality shale, gas locked in impervious shales and sandstones, methane found in coal beds or dissolved in brines under great pressure at depths of 15,000 feet are all a part of our physical resource base. They can and should be accounted for in any total resource inventory, but they cannot and should not be considered comparable in reserves or subeconomic resources. The likelihood of their soon becoming producible under present or near-future prices and technology is small enough that their importance for present generations is uncertain. Thus considerable caution must be taken to avoid giving them too much leverage in current decisions. After 50 years of effort and anticipation, the first commercial barrel of U.S. shale oil has not yet been produced. To be deceived by a too hasty reliance on methane dissolved in the waters of the Gulf of Mexico would be foolish indeed.

Although the oil shales of the West have become the classic example of a "just-around-the-corner" resource, we must somehow account for such a vast quantity of hydrocarbons. Many oil and gas resource appraisals do not include the oil shales because they are restricted to conventional crude oil, natural gas, and natural gas liquids produced from wells. Other analysts do not include them because they are not economically producible at the present time. If, however, a complete accounting is desired, then it is appropriate to at least identify these as other occurrences or noneconomic resources that are currently not produced and are likely to be significantly more costly than other forms of energy now being used. Whether quantification is attempted depends upon the purposes of the inventory.

Hydrocarbons occur in many forms in nature. Just as there are many types of coal (anthracite, bituminous, or lignite), there are heavy oils, tar sands, and kerogens that will not flow to drilled wells. This requires the extraction of the material either through the use of heat and chemicals or physically mining the rock so that it can be processed above ground. Since these are sedimentary deposits, they can be vast in extent but highly variable in recoverable energy content. In effect, they are low-grade deposits requiring expensive processing. As such they must be considered as either subeconomic or probably nonexploitable in any period of time that is of significance to present generations.

A number of largely unexploited sources of methane, the most abundant of the natural gases, have also attracted considerable attention. Among these are natural gas in dense sandstones of the West and Devonian shales of the East where the rock is relatively impermeable and does not allow the gas to flow freely to a well. As a consequence, the drilling of a well in these formations is not often rewarded with a great quantity of producible gas or a high daily rate of production. Methane in coal is well-known as a hazard to mining and is actually recoverable by drilling holes in the coal bed in advance of mining. Another recent discovery has been of the presence of methane in underground salt water found at considerable depth in the Gulf Coast area. The gas is held in the water by the great pressures that exist at the depth.

Relatively simple calculations of the volume of oil shale in the Piceance basin, tarlike substances in Utah, and methane in coal beds or other geological settings yield vast quantities of energy that physically exist. However, like exotic rocks on the moon, the fact of their existence should not be confused with economic and technological accessibility.

In the other occurrence classification, there is also that portion of conventional oil and gas that we do not expect to recover. Similar to low-grade oil shale, it would be physically possible to produce this oil and gas at great cost. One could literally mine an oil reservoir and produce all of the oil, or let a gas well produce until there was no more pressure left. Obviously, long before this, it would be far more sensible to use some other source of energy. Thus, those portions of our oil and gas resources unlikely to ever be recoverable can be accounted for among the other occurrences.

PRODUCTIVE CAPACITY

It is virtually impossible to determine how much oil and gas can be produced in any given year solely on the basis of knowing the quantity of proved reserves of oil and gas. If the nation finds itself lining up at gas stations or shutting down factories because adequate pressure cannot be maintained in all the utility mains during cold weather, the immediate supply problem is the productive capability of the delivery system, not proved reserves or undiscovered resources.

Over the years little study has been directed toward understanding the limits of the delivery system upon which we depend to move energy from the well to the burner tip. For the fossil fuel group, we have only the American Petroleum Institute's (API) estimate of productive capacity. This is the maximum daily rate of production that could be attained under specific condition on March 31 of any given year. It would require 90 days to reach, starting January 1, and is based upon existing wells, well equipment, and surface facilities. The estimate provides for no reduction in ultimate recovery, and environmental damage or other hazards are not accepted.

Obviously, it is useful to have such a measure of our capability. It is important, however, to be aware that the API definition of national productive capacity does not imply anything about the sustainability of this capacity over any specified period of time beyond the 90 days. Before March 31 of the year of estimate, the productive capability would begin to decline. This particular measure of productive capacity does not encompass our capability, or lack of it, for storage, transportation, and processing facilities to handle the oil once produced.

The inability to discriminate between reserves and deliverability is the source of considerable confusion in the reporting on the oil and gas situation in the United States. References to the estimated total reserves in a field or resources in a new region are equated with annual production or requirements. Since only approximately 10 to 15 percent of the reserves of a field can be produced in a single year, if connected to a delivery system, billions of barrels of oil reserves or trillions of cubic feet of gas translate into a much smaller amount of oil and gas available even in the early peak years.

To understand oil and gas supply requires more than a realization that estimating reserves is an inexact process; it also involves an appreciation of the limited capability of a well to produce oil and gas upon demand. The time required to explore, find, drill development wells, lay pipe, and provide process facilities is a further restraint on translating reserves into production. To this year's energy consumer the only supply that counts is deliverable oil and gas, not reserves or resources. If that flow is inadequate, periods of five to ten years or more and considerable investment will be required to alter it in any significant way. Considerably more attention to the limits of this process, and perhaps less to reserves, seems warranted.

It is understandable that the productive capacity of the vast oil-and gas-producing industry and its downstream facilities presents problems in terms of measurement. One would expect, in contrast, that the capability of a known producing reservoir would be a reasonably precise number. This has taken on a new importance in recent years, as questions have been asked about whether or not producers holding federal leases have been producing oil and gas as diligently as they might if the price for oil and gas were higher.

CONCLUSIONS

The Resources for the Future (RFF) staff has now engaged in over two years of studying, discussing, and explaining the uncertainties of oil and resource estimation. That has led not to better numbers, but to perhaps a better understanding of what the existing data can or cannot do for us. By and large, we find that most examinations of oil and gas resources reflect in part the professional background of the estimator but most importantly the purpose for which the work is designed. Many estimates that have been published provide total future quantities of oil and gas that may be produced rather than supply in the economic sense or rates of production over future time. All too frequently, these totals may be translated into years of remaining oil and gas by dividing them by the current or some other assumed rate of production. This leads to the too-simple conclusion that we may be out of oil and gas at the end of that number of years.

Published estimates of total future producible hydrocarbon fluids provide the public with a narrow view of future oil and gas supply. This is compounded by the fact that the public does not know how to interpret the figures. As one RFF workshop participant noted, "the difficulty [in publishing estimates] was the problem and confusion in the public's mind of what all these numbers mean. It has just been an absolute mess. They [the public] have taken undiscovered resources and related them to reserves, and this was not really our intent at all. Suddenly, we find ourselves quoted in the most peculiar ways. And much to our embarrassment."

To the economist, it is important to know how much and for what period of time oil and gas production rates might be increased by a change in the economic structure of the industry or in the cost-price ratios. Or how sensitive future oil and gas production rates are to changes in technology. Answers to these kinds of questions are not contained in the usual published estimates or future oil and gas reserves. As a participant in one RFF workshop said, "Not a single technique, approach, publication, or anything, has yet adequately dealt with what the economists would call the supply schedule. Somehow we have got to get some indication of what different levels of future supply would be available at different cost-price relationships."

In any given year the production of U.S. crude oil encompasses production from a reservoir that has been newly discovered, along with the production from reservoirs 10 to 50 years of age. The important point to note is that the future rate of supply will be a composite of the rates from both old and new reservoirs.

The experts who have appeared in the RFF workshops agree that the rates of oil and gas production cannot be increased indefinitely. At some point, the rates must inevitably peak and then continue to decline until the hydrocarbon resources of the search are exhausted. What the experts cannot agree upon is whether the annual rates of oil and gas supply can still be increased, and over what period of time, before the final decline begins. The extremes are illustrated by proposing different extensions in the future of the past oil production rate curve. The conservative view states that most of our giant fields have been found; the maximum annual rate has already been reached; and that, henceforth, there will be nothing but decline. An optimistic view might be that the annual oil production rate may still rise in response to exploration and new technology to some future maximum from whence it would decline until all reservoirs were exhausted. A moderate view would be a future annual production rate curve somewhat between these two extremes.

Whatever extension is predicted fur the future rate of oil supply, the area under the resulting profile of the future can be no greater than the total amount of oil that one estimates can ultimately be recovered. Thus, the estimator of the conservative situations not only envisions a declining rate of production but also a limited amount of total production yet to be achieved. The optimistic estimator sees not only an increased annual rate of supply but also a larger volume of oil yet to be produced. Whether the estimator approaches the problem in terms of rates or total future production, the results of the estimate must be internally consistent with respect to the relationship between rates and cumulative production.

Many of the published estimates of future oil and gas supplies have provided a value for the total supply through use of the traditional volumetric method. No matter how polished and sophisticated the details may be, the volumetric method still contains two basic perceptions: first, that the occurrence of hydrocarbons in an unexplored geological region and the parameters that are associated with its occurrence will probably bear a relationship to previously explored regions; and second, that the searching efficiency for finding oil in new areas will likely resemble what it has been in the past in older regions.

Any resource estimate for future oil and gas production in an unknown region may convey a misleading impression to the nonprofessional. The only thing an estimator can say with absolute accuracy is that he does not know whether there is oil or gas in a given region until wells are drilled to find out. The history of past estimates is rife with situations in which either little oil was found in areas where there was a high expectancy or much oil was found in areas where there was a low expectancy.

If one prefers to approach the future by considering the production rate curve rather than total future production, then the immediate problem is the lack of an established theoretical basis for predicting the future shape of that curve. It appears to be somewhat unwarranted to assume that the curve would be symmetrical. On the contrary, there is reason to suggest that the rising leg of the curve is dominated by one physical and economic process, that is, the discovery of new reservoirs, while the decline will be dominated by a different physical and economic process, that is, the depletion of discovered reservoirs.

The nature of oil and gas reservoirs is such that the highest rates of production occur in the early life of the reservoir. The flow factors of a reservoir taken together generally mean that more than half the production from a particular reservoir will occur after the maximum rate of production is reached. If it is possible to supplement the natural producing energy of the reservoir or to apply technology that will make more of the reservoir oil accessible to production, the history of the reservoir may show additional production rate peaks after the first one has been reached. This was the case in Pennsylvania and in Illinois.

The best current estimate is that, with present technology and prices, an average of between 30 and 40 percent of the oil known to exist in discovered reservoirs will have been produced at the time the reservoir is considered commercially exhausted. This is an average to be interpreted as we understand expectancy, that is, some people die younger and some live longer. The amount that can be taken from a particular reservoir is dependent upon the nature of the oil itself and the nature of the reservoir. There are known oil reservoirs from which the ultimate production will be as much as 80 percent of the oil contained in the reservoir. In other instances, the amount of oil that can be produced with present technology and prices may be as low as 10 or 15 percent of the oil in the reservoir. If technology improvement or a price rise permitted an abrupt change in recovery factors, a late production rate peak might show in the oil supply curve.

Another approach to examining the future of oil and gas is an examination of the rate at which exploratory drilling finds new oil and gas reservoirs. This approach does not necessarily depend upon prior estimate of whether oil or gas is present. It assumes that, if oil and gas are present, they will be found. The approach requires, however, an estimate of the efficiency for finding in the future. The published graphs that show the manner in which the amount of new oil found in the past related to the total exploratory footage drilled indicate that the finding rate has been decreasing. The reasons generally given for why less oil is being found per foot of exploratory hole drilled include:

- We have already found the big reservoirs and those near the surface. Future reservoirs will be found at deeper horizons.
- The most desirable geological regions have been explored and drilled.
- The geological areas remaining to be drilled are more inaccessible and more expensive; for example, on the continental shelf.

Whatever the reasons may be for an expected decline in search efficiency, it is relatively easy to see that if one extrapolates this decline into the future, the contribution to production rates due to finding new reservoirs will diminish. Consequently, one would conclude that the maximum production rate probably has been reached, but not all the experts agree that the search efficiency must decline. Both technology and economics could affect the trend.

Disagreement resulting from various methods of estimating future oil and gas supply revolves around whether the volume of oil and gas remaining to be found or our productive capability is the primary limiting factor on the domestic production rate in the immediate years ahead. The National Petroleum Council concluded from its inquiry that at least until 1985 the amount remaining to be found is not the limiting factor. NPC visualizes that annual production can increase with appropriate attention to the drilling rates, finding rates, improvements in recovery factor, and economic adjustments.

The straightforward production-history approach of M. King Hubbert and others, which is appealing to many, does appear to be very useful in telling us what is likely to happen in the near future if we continue doing things more or less the way we have been. It implies that production, drilling, and so on are insensitive to economics and policies. Barring an almost total interruption in exploration, such curves do seem to provide us with what should be our minimum expectation for future U. S. oil and gas output. However, this is not to suggest that any of the other kinds of appraisals are totally free of ties to past reserve and in-place figures and historical, economic, and technological factors.

There is greater satisfaction with recent estimates of future oil and gas supplies because the numbers appear to be converging. Instead of difference in orders of magnitude, two estimates may be within 10 or 25 percent of one another. In part, this merely reflects a greater consistency in methodology and assumptions than previously. Any comfort derived from this apparent consensus can be false. Although two estimators may now agree, even if they have used different methods, this does not necessarily mean that they are both right.

Finally, it is important to emphasize that all oil and gas resource estimates by the many analyses both public and private are dependent upon the same sets of numbers as starting points. Beyond this, there is no right methodology, and estimates are sophisticated guesses at best. All experts agree that the usable oil and gas hydrocarbon resources are probably sufficiently limited that the maximum annual rate of production and the decline until reserves are exhausted are events that will fall within a few decades, not much beyond that. The peak in the United States may have already been reached. Yet one must not minimize the importance of capturing the remaining one-half or one-third of our oil and gas.

So basically we are dealing with forecasts of annual production rates for two or three decades, and we must get an idea of the impacts of economic and technological changes on these rates. Cost data should be assembled so that it will be possible to analyze better the responses to economic change. Attention must be directed to the effect of technological change on increasing the recovery factor of existing reservoirs and the lead times needed to accomplish this.

This work will be aided if we can eliminate some of the past disagreements of estimators that stemmed from a lack of consistency in defining recovery factors and other concepts employed. If estimators are agreed on anything, it is that the definitions of terms must be examined closely and more standard definitions accepted for future resolution, if not of the supply question, then at least of why the estimators disagree in fact. Such a resolution would be an important step toward substantive agreement.

This is not as easy as it may seem. Even a seemingly simple term "total oil and gas in place" changes in meaning due to changes in information and the economic or technological perception of the analyst. Gas in tight sands or heavy oils would not have been encompassed within the definition of that term a few years ago. Yet some output from these sources has now joined the production stream.

3

THE GEOPOLITICS OF OIL

Stansfield Turner

In 1977 the Central Intelligence Agency (CIA) produced its first unclassified report of the international energy outlook. At the time, the report was roundly criticized as being overly pessimistic. As things turned out the CIA was not pessimistic enough. The events of 1979 once again demonstrated that the energy problem is a serious one. Moreover, the energy situation has substantial economic, political, and military implications as events in the Mideast and Afghanistan have demonstrated.

WORLD OIL OUTLOOK

World oil production is probably at or near its peak and will decline throughout the 1980s. Leaving aside unforeseeable supply disruptions, such as may result from political events or natural catastrophes, the rate of decline will depend on the success of oil exploration efforts in locating new reserves and on production policies in key oil-exporting countries. The CIA expects that output in the

---

At the California and World Oil: The Strategic Horizon Conference held at the University of Southern California in June 1980, John Eckland presented the Central Intelligence Agency's assessment of world oil supplies, summarized in this excerpt from the testimony of CIA Director Admiral Stansfield Turner before the U.S. Senate Committee on Energy and Natural Resources, April 22, 1980. Eckland is chief of the CIA's Petroleum Supply Analysis Center.

Persian Gulf countries will at best remain near current levels; production in other OPEC countries will fall; the Organization for Economic Cooperation and Development (OECD) output will begin to drop after the mid-1980s (including both our North Slope and the North Sea); and Communist countries will shift from a net export to a net import position in oil. Production will rise in the developing nations group, the size of the increase largely hinging on Mexican output decisions.

Simply put, the expected decline in oil production is the result of a rapid exhaustion of accessible deposits of conventional crude oil. In the 1970s, total new oil discoveries were probably no more than half as large as depletion. It is doubted that new oil discoveries in the decade of the 1980s can reverse this trend.

There is good reason to believe that the most prolific oil-producing areas have already been located and drilled. Even with modern technology, the chance of finding new giant fields is diminishing. Areas with the highest potential include the continental shelf of Argentina, the South China Sea, and some Arctic regions, partly in the USSR. Other areas thought to have good oil potential—such as offshore Burma and eastern Peru—have so far yielded disappointing results. Exploration efforts in the South China Sea and in the Gulf of Thailand are being hampered by conflicting territorial claims. Moreover, lead times are such that the chances of finding and producing large quantities before 1990 in the new areas are slim indeed.

The present high rate of depletion is causing producing countries to revamp their oil production policies. Moreover, during the past decade national governments have come to control the bulk of world oil resources. They take a longer view than did the private oil companies and see oil left in the ground as a better investment than excess oil revenues placed in foreign money markets. Many of these governments are disenchanted by the idea of overheating their economies through rapid modernization programs, the large presence of foreign workers, and the social strains that accompany large inflows of oil wealth. Major producers like Saudi Arabia, Kuwait, and Iraq can cover their development needs with far lower than current rates of oil production.

The projected decline in oil supplies therefore results from both a technical analysis of output in countries that are now constrained by declining resources and from a political assessment of production policies in countries constrained by their own policy decisions. The CIA projections at the end of the 1970s were:

- Among the OPEC group, inadequate reserves will cause production to drop during the 1980s in Venezuela, Nigeria, Ecuador, and Gabon, barring an unforeseen turnaround in exploration results. The decline in Venezuela could be slowed by the development of heavy oil deposits and new offshore oil fields.

- Oil output will stagnate in Canada; the projected oil yield from tar sands will only offset declining production of conventional crude. Some of the northern areas are geologically promising but the odds on finding gas are greater than on finding oil; in any event northern oil as well as recently discovered oil off the Newfoundland coast probably could not be delivered on a large scale in the 1980s.
- Production in the United States probably will continue to decline despite heavy drilling activity; most U.S. companies in 1979 reduced their projections of output in the 1980s.
- Production in the USSR was expected to peak and then to decline steadily throughout the decade. The Soviet estimate will be dealt with separately because it is a key factor in the overall political and economic assessment of the energy scene.
- Despite limited reserves, production should increase in China if Western technological assistance is forthcoming. Chinese consumption, however, will increase rapidly, absorbing most of the rise in output.

The major countries with reserves sufficient to support a boost in output are Saudi Arabia, Kuwait, Iraq, Iran, the United Arab Emirates, Norway, the United Kingdom, and Mexico:

- Saudi Arabia may increase its capacity 1 million b/d or so above its present 9.5 million b/d level. But the Saudis may reduce actual output to 8.5 million b/d or lower.
- Kuwait could maintain output at its current capacity of 2.7 million b/d for at least 50 years, but it has reduced output to 1.5 million b/d because of an inability to productively use its oil revenues.
- Iraq would like to raise capacity in order to increase its influence in OPEC and Middle East circles. Production, however, will probably be held to the level needed to meet development needs—about 3 million to 3.5 million b/d or less, as oil prices rise.
- Iran under the current leadership has opted for slow economic development that can be sustained by an oil output of only 2.5 million to 3 million b/d. Even this level of output will require foreign assistance and large investments to put the oil fields back in shape.
- Output in the United Arab Emirates, Qatar, and the Saudi-

Kuwaiti Neutral Zone is liable to decline slowly during this decade. Only Abu Dhabi has a recognized potential to add capacity but water encroachment problems will limit output.

- North Sea resources are sufficient to support some gain in production during the next few years but both the British and Norwegian governments have adopted long-term conservation policies.
- Mexico will provide the only major output increase in the 1980s. Producing 2 million b/d in 1980, the government is rapidly increasing output to meet revenue needs. Although a level of 4 million b/d or more probably could be achieved and sustained later in this decade, economic bottlenecks and inflation worries may make it undesirable to go that high.
- Production from other countries will not have a significant impact on world oil supplies in the 1980s.

Under optimistic assumptions then, free world oil supplies will decline slowly if there are no major disruptions. On the other hand, OPEC production cuts for whatever motive could mean a substantial drop in supplies over the decade. In either case, OECD countries are virtually certain to get a declining share of available oil. OPEC members and other developing country oil producers will increase their own consumption and the faster growing LDCs are likely to get enough preferential treatment from OPEC to increase their share of the pie.

TOTAL ENERGY SUPPLIES

In this decade at least, nonoil energy sources are not likely to offset the projected slippage in oil production. The prospect for continued increases in oil prices does create strong incentives to develop nonoil resources but development of coal, nuclear power, and hydroelectric resources continues to be hampered by environmental and/or safety concerns. The 8–10 year lead time for nuclear power plants would in any case delay a rapid buildup in capacity until the 1990s. Except for France, more nuclear plants have been cancelled than ordered by Western countries in the past four years.

Natural gas supplies will increase only slightly, if at all, during the 1980s. Declining U.S. output will be offset by increased availability of Soviet and Mexican pipeline gas, some rise in West European and Canadian output, and a boost in LNG exports, mainly from Algeria. U.S. output could fall sharply during this decade unless faster progress is made in finding new reserves and in developing technology to exploit hard-to-reach reservoirs.

At best, total energy supplies for the OECD countries will increase slowly during the 1980s, but total energy supply could also fall.

In either case, the energy outlook presents serious economic problems. To achieve acceptable rates of economic growth, fuel conservation will have to increase dramatically. If energy supplies remain constant, achieving a 3 percent economic growth rate would require that by 1990 the rate of fuel consumption per unit of GNP be reduced by one-third. While price increases are clearly spurring such conservation, slower economic growth will probably entail a slower growth in investment in capital plant and this in turn will reduce the potential rate at which new energy-efficient techniques can be introduced.

THE SOVIET SITUATION

The OECD outlook is complicated by what we see happening to Soviet oil output in this decade. The USSR was the world's leading oil producer with a output of 11.7 million barrels per day in 1979. Production growth, however, has slowed markedly in recent years. Output was expected to peak in 1980 at less than 12 million b/d and begin falling the next year.

Given this Soviet production outlook, the Communist countries as a group were projected to shift from a net export position of 0.8 million b/d in 1979 to a net import position of at least 1 million b/d in 1985. It should be noted that this is a conservative estimate of what these countries could afford to pay. Unconstrained demand for imports would require substantially larger net imports.

Oil production is already declining in all of the major oil-producing regions except West Siberia, and further gains even there are uncertain now that the supergiant Samotlor oil field has reached a production plateau. Samotlor, which has accounted for the bulk of production growth in recent years, is likely to begin to slump in the early 1980s and then fall rapidly. Meanwhile, the decline already underway in older major producing regions probably will accelerate as reserves are depleted.

By 1985, Soviet oil output probably will have slid to 10 million b/d or less. The 10 million b/d figure assumes that exploration is relatively successful, development drilling goes well, and the Soviets can acquire the needed equipment and technology, mainly from the West. In the last half of the 1980s, production probably will continue to decline but at a slower rate.

In the longer run, the future of Soviet oil production depends on Soviet success in discovering and developing oil fields in new areas—primarily in the Barents and Kara Seas, the Caspian Sea, Eastern

Siberia, and the deep onshore Caspian depression. These areas are only beginning to be explored so that any new finds would have little impact on oil production until the late 1980s or early 1990s. Moreover, much of the technology for exploring and developing resources in the Barents and Kara Seas is not even available in the West. Similarly, the Soviets as yet have not found a way to extract their large known reserves of heavy oil, much of which is nonflowing.

The prospective decline in Soviet oil production is only part of the problem. Coal output is nearly stagnant, and natural gas production, although growing rapidly, is limited by the capacity to increase pipeline beyond the current all-out effort. Meanwhile, Soviet savings of energy through conservation have been and will continue to be small. The Soviets are already comparatively efficient in heat production, electricity generation, and rail transport. Future gains in these areas would require a costly replacement of existing capital stock.

Leonid Brezhnev told President Carter at the summit that energy was Moscow's most pressing problem. However, there is no clear consensus among Soviet leaders on what the energy situation portends.

Soviet leaders will probably try to weather the crunch with a combination of belt-tightening, slower economic growth, cutbacks in oil exports, and increased oil imports on the best terms they can get. While the exact mix of adjustments is unpredictable, much of it will be slower economic growth, which means smaller increments in goods and services each year to divide among defense spending, consumer welfare, and investment. Since defense spending as well as industrial investment probably will continue to take priority, the big loser inevitably will be the Soviet consumer. The consumer's position could deteriorate suddenly and sharply if a bad agricultural year coincides with the onset of the oil crunch—the Soviets have been averaging one bad crop year in three during the last two decades.

Because Eastern Europe depends so heavily on Soviet energy supplies and cannot afford to buy much oil elsewhere, falling Soviet oil production will curtail growth in Eastern Europe and raise the prospect of stagnating living standards. These dislocations will put a good deal of strain both on the Kremlin leadership and on the Warsaw Pact alliance. Relations within the pact are already uneasy because of the Soviet adventure in Afghanistan. Given the advanced age of Soviet leaders, the oil crunch is likely to occur during a large-scale changeover in the Soviet Politburo.

The policy options open to the Soviet leadership to adjust to reduced oil supplies are all extremely painful. To ease the problem, Moscow no doubt will make an intense effort to obtain oil at concessionary prices from the oil-producing countries through barter deals, sometimes involving arms sales. More forceful action, ranging from covert subversion to intimidation, or, in the extreme, military action, cannot be ruled out.

IMPACT ON THE MIDDLE EAST

Taken together, the Western and Soviet oil outlook sets the stage for an East-West competition for Mideast oil, adding another potentially destabilizing ingredient to an area that already has experienced wrenching political events in the past year or so. For example:

- Existing Middle East alliance patterns have been disrupted by reactions to the Egyptian-Israeli Peace Accords and by the overthrow of the shah of Iran. Both events have led Iraq to attempt to assume a position of leadership in the Arab world.
- Soviet involvement in the Middle East has also rebounded in recent years. A friendship treaty with South Yemen and an arms agreement with North Yemen gave Moscow a firmer foothold on the Arabian peninsula. The occupation of Afghanistan put them in a better position to intimidate Pakistan and Iran, so long as the Soviets could bring the Afghan insurgents under some sort of control.
- The influx of oil wealth with its accompanying modernization programs and massive labor movements has aggravated internal social and ethnic and religious strains, most obviously in Iran but also in other Persian Gulf countries.

In the longer run, ethnic and religious cleavages in the Middle East could flare into conflicts that would threaten oil supplies. This analysis can discern the warning signs although the exact flash points will continue to be extremely hard to pinpoint.

It also is likely that the Soviets will be increasingly active in the diplomatic arena in the Middle East, holding out as a carrot the glimmer of a stable political atmosphere if the Gulf states become more cooperative on oil and political matters. Moscow already is strongly making the point that Middle Eastern oil is not the exclusive preserve of the West, most recently in a suggestion to "include the security of the oil routes" in a Soviet-proposed European conference on energy.

IMPACT ON WESTERN RELATIONS

Competition for available oil supplies not only will put added pressure on East-West relations, it will strain relations among the

industrialized Western nations themselves. Western energy cooperation has worked fairly well so far but current IEA and EC arrangements to share available oil in the event of an emergency affecting
one or more members have not been put to the test. The going will
get tougher as the decade progresses and as the IEA oil import targets adopted for 1985 have to be substantially scaled down.

Most major foreign governments have essentially left it to
higher prices to reduce oil demand and imports. France has done the
most with administrative controls and incentives to discourage energy
use and promote the substitution of nonoil energy sources, especially
nuclear power. Japan and Italy have drafted conservation programs
that have yet to be enacted into law; Great Britain has a stand-by
conservation plan; and aside from mandatory heating limits for government buildings, the West Germans have confined their program
to a public energy-saving information campaign and voluntary plans
for industry and transportation.

The question of energy supply has had, and probably will continue to have, a more important impact on Western external relationships than on internal policy. Realizing that there is not a great deal
of scope over the next decade or so to diversify their fuel sources,
the West Europeans and Japanese have been trying various means to
improve their standing with the Middle Eastern oil producers that
include the following:

- Proposals are still on the table for a Euro-Arab dialogue on
  energy and other economic issues. Thus far, this has been
  an unsuccessful attempt to forge a special political-economic
  relationship between Europe and the Gulf states.
- Efforts to secure state-to-state oil contracts with the Gulf
  producers continue. The largest arrangement is a recent
  West German-Saudi medium-term contract for 300,000 b/d
  annually, although most deals have been for smaller amounts.

Earlier the Europeans adopted a passive policy, avoiding state-
ments and actions that would risk disruption of imports from the Gulf.
That policy has given way to a more active attempt to ensure supplies.
Direct European involvement in Middle Eastern issues extends to in-
creasing concern with the Egyptian-Israeli discussions on West Bank
autonomy. An impasse in the negotiations probably would bring forth
European proposals for reviving talks under United Nations auspices.
These proposals probably would reflect a more pro-Palestinian, pro-
Arab stance than the Europeans have so far taken.

The Europeans, of course, are sensitive to Soviet activities in
the Middle East. They immediately rejected the 1980 Soviet overture
for talks on Gulf security, but they will want to keep lines of communication open to Moscow to try to diffuse any conflicts that might

threaten oil supplies from the Gulf or lead to confrontation with the
Warsaw Pact.

IMPLICATIONS FOR THE UNITED STATES

The peaking of world oil output and the competition for available
supplies by the West, the LDCs, and soon the Eastern bloc all adds
up to less Middle Eastern oil for the United States. At the same time,
the Middle East remains critical to U.S. interests—we will still need
whatever oil we can get from the area and our allies will remain over-
whelmingly dependent on these sources.

All of the obstacles to securing a stable flow of oil from the
Gulf will be magnified during this decade. The physical security of
the oil routes and any of the Gulf oil fields is tenuous and any major
intraregional conflicts—such as Iran or another Arab-Israeli war—
could well lead to some disruption of oil supplies. Equally important,
in the view of friendly governments in the Gulf, our ability to offer
protection in the area has been seriously eroded. Yet most of these
states still fear that a substantial U.S. presence in the Gulf would be
destabilizing politically. To protect themselves during a period of
heightened superpower competition, most will be inclined to distance
themselves enough from the United States, or in a few cases from the
USSR, in order to build a more neutral image on East-West issues
rather than to clearly choose sides.

In summary, the world faces a decline in oil consumption.
Neither efforts to find more oil nor to provide fuel substitutes can
compensate for the shortfall during the next ten years. Greatly accelerated energy conservation will be necessary but difficult to achieve,
with much of the burden falling on the United States as by far the
largest per capita consumer. Economic growth is likely to be slower
in the 1980s than in the 1970s. Although many factors other than
energy supply will affect economic performance, the ability to increase GNP with lesser accompanying increases in energy consumption will be critically important.

Politically, the cardinal issue is how vicious the struggle for
energy supplies will become. This competition will create a severe
test of the cohesiveness of both the Western and Eastern alliances.
Developing forms of cooperation among oil-consuming countries to
regulate this competition and to prevent it from becoming mutually
destructive will constitute a critical challenge. The entrance of the
Soviet Union into the free world's competition for oil not only further
squeezes oil supplies available to the West but also entails major
security risks.

III

The Middle East and
U. S. Oil Supplies

INTRODUCTION

Whereas the previous two parts focused on world energy supplies from a broad worldwide perspective, the chapters that follow in this part deal primarily with the Middle East region of the world. The impact on world energy supplies of just six Middle Eastern members of OPEC is evident in that the six nations at the start of the present decade controlled over 50 percent of the world's proven oil reserves, produced one-third of the world's oil, and supplied approximately 62 percent of the oil traded internationally. In Chapter 4, Fereidun Fesharaki describes the crucial role played by the nations located in the Persian Gulf area in the world oil market, in OPEC, and possibly in the economic and political security of the world. For the United States and other nations, this region is vital. However, as Fesharaki points out, the climate in the 1980s is unlike that of earlier periods; nations and governments and not private companies set production levels. As a result, oil is seen as an exhaustible resource and not one to be depleted simply because consumer nations want a lower price. In short, production will not increase and prices will not decrease.

A somewhat different perspective on the impact of the region on the United States is provided by Tom E. Burns in Chapter 5. He seems to offer a more optimistic view of the potential contributions of nonoil resources, but also concludes that the United States will remain dependent on foreign oil, including that produced by OPEC nations. Burns also forecasts a higher level of production for OPEC and Middle Eastern nations, but his conclusion remains the same: dependence on foreign oil supplies throughout this century.

Finally, Thomas L. Neff adds another dimension to the discussion by delineating the changes that have occurred in producer nations and in the world oil market. Neff argues that understanding market realities is crucial to the development of appropriate energy security policies. An underlying fact of the situation in the 1980s and beyond is that the world oil market is far less flexible and resistant to disruption than previously was the case. As such, the United States and other nations must pursue measures to increase flexibility and in doing so, the costs will be high both in economic and political terms.

Although these contributors paint a picture that would appear at first glance to be on the pessimistic side, each also offers suggestions on how the United States and other nations can cope with the changing world oil market. However, it also is clear that simplistic approaches to the problems will undoubtedly fail.

4

THE FUTURE OF OPEC POLICY

Fereidun Fesharaki

INTRODUCTION

In the early 1970s, after a decade of quiet, ineffective existence, a combination of factors led to the emergence of the Organization of Petroleum Exporting Countries (OPEC) as a major actor in the world petroleum market.

For many, the emergence of OPEC meant higher petroleum prices. OPEC was seen as a cartel of greedy, self-serving nations who tried to extract all the profits they could from the petroleum-dependent world. Most analytical work in the early part of the 1970s was confined to finding a means of breaking up OPEC. OPEC was seen as a serious threat to the world's economic and political security. Western analysts were puzzled how an organization comprised of nations with such social, political, and economic differences could get off the ground and continue to survive. Eminent academicians suggested various means of encouraging one member nation to desert the others. But these analysts failed to realize that the change in the oil market power balance and the assertion of sovereign nations' rights over their oil resources are not temporary phenomena but are here to stay. Oil-consuming nations wasted much valuable time that could have been spent seeking alternative energy sources and, perhaps, reaching an agreement with the OPEC producers.

As the legitimacy of the OPEC nations' actions came to be generally recognized, the 1973-74 price explosion clearly showed that the era of cheap oil was permanently over. Significant actions were called for to diversify the world's energy supply. The industrial nations of the world realized their urgent need to reduce their dependence on imported petroleum.

But the 1975 recession, partly caused by the oil price increases (from $1.80 per barrel in 1970 to $12.65 per barrel in 1975[1]), weakened the demand for the price of oil. In 1975, OPEC's oil production fell by 11 percent; thereafter, nominal prices increased at rates below the level of world inflation. As a result, the real price of oil fell for four straight years. A recent study examining the impact of dollar devaluation and inflation on the cost of crude shows that for Japan and West Germany the real cost of crude had declined by around 20 percent in the four-year period.

Only by the third quarter of 1979 had the oil price increases fully compensated for the declines in the preceding years. At a real cost of $7-11 per barrel in the third quarter of 1979, the price of oil reached the 1974 level in real terms.[2] Also, the large OPEC trade surplus, $60 billion in 1974, had by the end of 1978 been reduced to only $1 billion.[3] The international banking system proved to be far more flexible than many had expected, successfully channeling the surplus capital to both industrialized and developing nations.

In retrospect, both the industrial and developing nations managed successfully to escape major economic problems despite rapid changes in the price of oil. The industrial economies grew at an average annual rate of 3.4 percent during 1970-78, while the rate of economic growth for the developing world was 5.7 percent for 1970-76 and 4.8 percent for 1977.[4]

Because of the declining real price of oil, diminishing OPEC surplus, and relatively successful world economic performance, by the end of 1978 the earlier sense of urgency surrounding petroleum supplies and prices had given way to a sense of complacency reminiscent of the earlier decades of cheap and abundant oil.

The February 1979 revolution in Iran reversed the trend of oil prices and again brought home the message that the problem of oil supplies and prices would not go away by itself.[5] In just one year— January 1979 to January 1980—oil prices increased by 120 percent to an average of $30 per barrel. The OPEC surplus of $1 billion in 1978 rose to around $50 billion in 1979 and exceeded $100 billion in 1980.[6]

Two distinct increases in oil prices have occurred: the first in 1973-74, the second in 1979. Though the percentage increase in the real price of oil was greater in the former than in the latter, it is likely that ramifications of the latter increase will be far more significant. In the future, the supply of petroleum may be a more important issue than the price of petroleum (although the two issues are intertwined). This chapter will consider the implications of the 1979 changes from OPEC's point of view and identify likely trends in the world oil market and OPEC policy.

FORECASTING OPEC PRODUCTION

No unified production policy exists within OPEC. In other words, there is no production prorating program as such. However, as will be explained later, member nations have acted in a way that precludes a need for prorating. Decisions and signals from different nations are interpreted by other exporters so that production is curtailed once a glut develops, perhaps with a six- to twelve-month lag.

Western analysts have traditionally forecasted OPEC production by assuming that OPEC would produce to fill the gap left by other producers in satisfying the world's energy demand. That is, analysts first project world energy demand, then subtract all nonoil sources of energy, further subtract non-OPEC oil production capacity, and take the remainder as OPEC's forecasted level of production.

This method of forecasting OPEC production is a poor one and is certain to be misleading. First, it does not take into consideration the amount of oil OPEC members prefer to produce, considering the economic, social, and political implications. What OPEC wants to produce may be different from what OPEC technically can produce or what the world needs. Second, it does not allow for the possibility that OPEC's policy determines the development of other energy sources, not vice versa.

The only way to reach rational projections of OPEC supplies is through the so-called "bean counting" method. That is, each member nation's production level must be estimated separately, according to that nation's production policy. The sum of these forecasts is the forecast for OPEC.

During the 1970s, many OPEC nations became convinced that they should conserve their oil resources. They are likely to continue their policy of decreased production and crude exports for several political and economic reasons.

First, governments, not oil companies, now set production levels. Thus, the calculus of relatively short-term profit maximization applicable to the case of private oil companies is not relevant.

Second, oil is an exhaustible resource whose depletion may throw the producing nations back to what they see as the dark ages of poverty and domination by industrial powers. The reserve/production ratio is falling and little possibility of additional large-scale oil discovery exists.

Third, the original hopes of rapid modernization and industrialization through the use of oil income have now faded away. It has become clear to OPEC members that the injection of oil money will not by itself bring about such development. Self-sustained growth is an evolutionary process; therefore, the life of the oil revenues must be extended to ensure a continuous flow of foreign exchange.

Fourth, OPEC countries that have invested surplus funds abroad have faced inflation and the declining value of the dollar. Furthermore, their invested assets are subject to seizure in the host country. The freeze of Iranian assets by the United States has had a major negative psychological impact on the oil exporters. Storing their wealth in the form of oil under the ground is far more attractive than many other investments.

Fifth, pressures from all sides of the political spectrum in many of these countries are mounting in favor of conservation. At the same time, rulers fear that excessive investments may disrupt the very fabric of social order in their countries, as in the case of Iran. Even conservative regimes are likely to cut production to appease the opposition. If pressure is strong enough, it could topple some regimes.

Sixth, OPEC nations do not think the industrial world has made sufficient progress in conserving oil. They feel the price has not gone high enough to force a serious effort to conserve. Thus, OPEC members refuse to deplete their dwindling asset to appease the appetites of industrial nations.

Seventh, domestic demand for oil products is rising rapidly. OPEC domestic demand rose from 770,000 barrels per day (b/d) in 1970 to around 2.2 million b/d by the end of the 1970s. It is expected to reach 3.9 million b/d in 1985, 6.3 million b/d in 1990, and 16.7 million b/d by the turn of the century.

Eighth, oil exporters realize that there will be no major substitutes for their oil in the next 10 to 20 years. They also know that supply reduction is the key to price increases. Reduced supplies, at the same time, given OPEC nations enormous political power that they can exercise to their own benefit and for the benefit of the other less-developed countries (LDCs).

Finally, by processing crudes domestically, oil exporters can obtain the value-added for themselves.

THE PERSIAN GULF PRODUCERS: PRODUCTION OUTLOOK

The six OPEC nations located in the Persian Gulf are the key to the future of the world oil market, OPEC, and perhaps the economic and political security of the world. In 1979, they controlled more than 50 percent of the world's proven reserves, produced one-third of the world's oil, and supplied around 62 percent of the oil traded internationally. Not only do they produce more than two-thirds of OPEC's oil, but they also have the largest potential for expanding production through both primary and secondary (enhanced recovery) techniques.

The Persian Gulf oil exporters have divergent foreign policies.

TABLE 4.1

1980 Anticipated OPEC Production Level (million b/d)

	Sustainable Capacity 1979[a]	Expected 1980 Production[b]	Decline from 1979[b]	Type of Restriction
Algeria	1.10–1.20	1.00	0.14	resource constrained
Ecuador	0.25	0.23	+0.02	resource constrained
Gabon	0.25	0.20	0.01	resource constrained
Venezuela	2.40	2.13	0.21	resource/policy constrained
Indonesia	1.65	1.56	0.04	resource constrained
Nigeria	2.40	2.15	0.18	resource/policy constrained
Libya	2.10	1.70	0.40	resource/policy constrained
Iran[c]	4.50	1.80	1.20	policy constrained
Iraq[d]	4.00	3.50	+0.10	policy constrained
Kuwait[e]	2.50	1.50	0.75	policy constrained
Saudi Arabia[f]	9.50–10.50	9.50	+0.30	policy constrained
United Arab Emirates	2.48	1.75	0.07	policy constrained
Qatar	0.65	0.48	0.03	resource constrained
Neutral Zone[g]	0.60	0.56	—	resource constrained
Total OPEC[h]	34.38–35.48	28.06	2.61	
Domestic consumption[i]	—	2.43		
Export availability	—	25.63	2.65	

116

[a] Capacity that can be maintained for several months. Does not necessarily reflect the maximum production rate without damage to the fields. Generally 90 to 95 percent of installed capacity.

[b] Based on official and unofficial declaration of production levels and available data compared to the first nine months of 1979. For countries where no declaration is made, the average 1980 production level for the first six months is assumed.

[c] Iran's capacity has declined from 7 million to 4.5 million b/d after the revolution, due to fall in pressure and lack of necessary maintenance. Iran's stated target for 1979 was 4 million b/d. In the second half of 1979 it produced between 3.5 million to 4 million b/d. Although no official announcement is made regarding 1980 production level, political and technical factors strongly suggest 2 million b/d cutback from 1979 target.

[d] Iraq has hinted a few times that it will reduce production by 600,000 b/d in 1980. Still the government is noncommittal, saying any cutback will depend on market developments. It is assumed here that in fact production will not be reduced in 1980.

[e] Kuwait has had a production ceiling of 2 million b/d for several years. It has declared its intention to reduce production to 1.5 million b/d.

[f] Saudi Arabia's previously anticipated expansion of reserves to 14 million to 20 million b/d has not materialized due to the Saudis' lack of desire to invest heavily in such expansion as well as some technical difficulties. The sustainable capacity is estimated by the CIA to be 9.5 million and Petroleum Intelligence Weekly to be 10.5 million b/d. However, more recent testing of capacity suggests a figure of 10.2 million b/d. The Saudi oil production ceiling of 8.5 million b/d was lifted for part of 1979 to cushion the impact of disruptions from Iran and reduce the rise in prices. It has been announced that production ceiling will not be operational for 1980. It is assumed here that production will be maintained in 1980 at 9.5 million b/d.

[g] Equally shared between Saudi Arabia and Kuwait.

[h] Based on estimates by the OPEC Secretariat. S. A. R. Kadhim and A. Al-Janabi, "Domestic Energy Requirements of OPEC Member Countries" (Paper presented at the OPEC seminar OPEC and Future Energy Markets, October 1979, forthcoming in OPEC Papers 1980).

[i] Data not available (—).

Source: Compiled by author, see note h. Printed in California and World Oil: The Strategic Horizon, The California Energy Commission, Sacramento, California, January 1981, p. 33.

Four nations—Saudi Arabia, Kuwait, the United Arab Emirates (UAE), and Qatar—are pro-Western. Iraq is nonaligned, but leans toward the Soviet Union and belongs to the radical Arab Rejectionist Front. Iran has no foreign policy posture as yet.

They also differ greatly internally. All but Saudi Arabia, Iraq, and Iran are tiny sheikdoms or city-states. Yet all six regimes are unstable and face internal opposition. Revolution, civil war, or major civil disobedience loom in their futures. Oil production, therefore, may be reduced as a matter of preferred government policy to pacify political opposition or because of political disorder.

Some OPEC nations produce as much oil as their resource base will permit, but those in the Persian Gulf have imposed production ceilings below their technical potential for policy reasons. Expected 1980 production levels are compared to 1979 levels in Table 4.1. Likely production from the key Persian Gulf states is discussed below.

Saudi Arabia. Saudi Arabia has a sustainable capacity of 10.5 million b/d and is producing 9.5 million b/d—1 million b/d above its self-imposed ceiling on production. Saudi Arabia's population is 7.5 million, of which 20 percent are guest labor. About one-quarter of the foreign workers are from the West; the others are Yemenites, Palestinians, and Lebanese. Around 1 million of the Saudis are Shiite Muslims—a minority in a nation of Sunni Muslims. However, the Shiite population is involved heavily in the oil fields and refineries. The Shiite Saudis together with Yemenite workers are politically active and oppose the royal family. Numerous instances of violence and street clashes have been reported since mid-1979. The Islamic fundamentalism of the Ayatollah Khomeini and leftist politics in South Yemen are behind the new wave of opposition. Little or no opposition comes from the Palestinian guest workers, who are well-paid technicians and technocrats.

Besides opposition from the underprivileged, there is also political opposition from the middle classes and even dissension within the royal family. Some powerful voices within the royal family favor a cutback in oil production to 4 million to 4.5 million b/d, among them the minister of planning, Hisham Nazar, and the minister of finance, Abal-Khail. Many Saudis are becoming less and less confident in the policy of producing oil to appease the United States and the rest of the industrial world. They feel the nation is producing oil far beyond its needs, flaring the valuable associated gas produced with oil, and investing the surplus funds abroad in depreciating U.S. dollars. The Saudis are getting little, if anything, in return. First, there has been little or no progress in the Arab/Palestinian/Israeli dispute. Second, the United States will not provide them with any guarantee such as indexation to protect Saudi investments in the United States. Third, the Saudis' request for more long-range fighter planes has been rejected because of pressure from the Israeli lobby in the United States. Fourth, some Saudis see as irrelevant apparent assurances that the United States will defend Saudi Arabia against external aggression. They believe that the country's problems arise from internal factions supported by external enemies, primarily the Soviet Union. The United States has not shown itself to be a reliable friend and, in any case, will not be able to quash internal uprisings against the regime.

The argument often advanced in the West that what the Saudis do is a cleverly designed course of action in their own long-term interest is false and misleading. According to this argument, the Saudis maintain high production levels and argue against price hikes in order to: first, slow down the development of alternative energy sources that could compete with Saudi oil in the next century, and second, protect the Western economies—particularly economically troubled nations such as Italy that could fall into Communist hands in the long run. The argument is fallacious because Saudi Arabia's decision-making process, like that of the other Middle Eastern countries, has a very short planning horizon. If the Saudis were worried about the twenty-first century, they would strive to loosen their autocratic hold over their nation, use the oil proceeds more carefully, and conserve their single valuable but exhaustible resource.

The Saudis are currently repeating many of the mistakes committed by the shah of Iran, particularly in terms of excessive injection of capital into their economy. It took only $20 billion annually of imports into Iran to overheat the economy, ruin the planning process, and break up the social fabric of the country, resulting in the overthrow of the regime. In 1980, Saudi Arabia was importing over $45 billion of goods and services (from an expected oil income of $90 billion). For a country commonly referred to by economists only three or four years ago as a "low absorber," this is bound to be a heavy shock. That, despite such inflow, the inflation rate is below 10 percent is indicative of the investments being highly capital intensive, technologically advanced, and located in enclaves or satellite cities outside the reach of the domestic economy. These investments, encouraged by the U.S. government to recycle the petrodollars, are going to require more and more foreign labor, particularly from the industrial nations. The continuing dualism in the economy will have very dangerous consequences.

In 1980 Saudi Arabia was expanding its oil production capacity to 12 million b/d. It is, however, highly unlikely that production will ever exceed the current level. The additional capacity is probably intended more for use in emergencies or to improve the Saudis' bargaining position within OPEC than actually to produce at that level.

In the future, the Saudis' actual production level will probably be determined by the $21 billion system for gathering natural gas, now under construction. Once complete, the system could utilize no more gas than would be associated with oil production of 6.3 million b/d. Any oil production over and above this level will require gas flaring. Because flaring is an obvious waste and a political embarrassment, Saudi Arabia will probably limit oil production to 6.3 million b/d beginning in the mid-1980s.

There have been suggestions that the United States could reach an agreement with Saudi Arabia through which the Saudis would continue to expand capacity to perhaps 15 million to 20 million b/d. In return, the United States would pledge to reduce its oil imports, relieving the pressure on the Saudis to maintain such high levels of production. The excess capacity would be used to keep the lid on prices and supply oil if there were an embargo. Such a proposition seems unworkable. Who would pay for costs associated with expansion and maintenance of capacity? And why should the Saudis do it? Is there anything that the United States could guarantee to do in return outside of the oil front? Given the structure of the United States political system, the answer must be no.

Iraq. In relative terms, Iraq is perhaps the most stable government in the region. Iraq's proven petroleum reserves, generally estimated in the oil literature at 31 billion barrels, are widely believed to be close to 100 billion barrels, second only to Saudi Arabia's. Of Iraq's population of 12 million, 60 percent are Shiite Muslims who constitute the underprivileged groups. The Sunni Muslim minority has ruled the country since the 1958 revolution. There is some internal opposition by the Shiite community and the Kurdish ethnic minority (who are Sunnis), but the socialist leadership seems to be in full control.

Iraq's very close relations with the Soviet Union have chilled since the late 1970s, particularly after the Soviet invasion of Afghanistan. Indeed, Iraq has sought to broaden its ties with the Western democracies since 1979.

Despite radical rhetoric, in reality, Iraq's oil policy has been very moderate. It did not effectively take part in the Arab oil embargo of 1973 on the grounds that the embargo did not go far enough. In 1975, when OPEC production declined by 11 percent, Iraq's output rose by 15 percent, indicating price discounts in back-to-back deals. [7] In 1979, Iraq, together with Kuwait and Venezuela, formed the moderate price group within OPEC. During both 1979 and 1980, Iraq maintained production at 3.5 million to 3.7 million barrels to replace lost exports from Iran.

Many analysts expect Iraq to emerge as the superpower of the Persian Gulf during the 1980s. Iraq is likely to step up oil production and expand capacity in its quest for influence and leadership in the

Persian Gulf. Unlike those in Iran and Saudi Arabia, Iraq's economic development programs have been relatively successful; there is little indication of oil-induced strains in the economic and social fabric of the society. Iraq will remain a moderating influence on the oil scene in the 1980s while remaining as nonaligned as a developing nation can be.

Iran. A nation of 35 million and the largest in the Persian Gulf, Iran is still suffering from postrevolutionary political upheavals that promise to keep the country unstable for the next few years. To maintain minimum rates of economic growth, oil production of 3 million b/d throughout the 1980s will be required. This is a level that one should expect whether the present regime survives or not. Technical problems such as the decline in reservoir pressure will not allow production far above 3.5 million to 4 million b/d in any case. A return to the 6 million b/d level reached during the shah's rule is not to be expected.

OPEC PRODUCTION FORECAST

For 1980-81, oil produced by the OPEC countries will be sufficient to meet world demand. But the oil glut of 1980 is unlikely to

TABLE 4.2

Likely Production and Export of OPEC Crude in the 1980s

	1980	1985 I	1985 II	1990 I	1990 II	2000 I	2000 II
Production	28.0 (47%)	28.0	25.9	28.0	23.6	28.0	?
Export	25.6 (83%)	24.1	22.0	21.7	17.3	11.3	?

Notes: Percentages are percent of world trade. Scenario I assumes constant 1980 output. Scenario II is based on Table 4.3. Production for the year 2000 is too uncertain to make a reasonable prediction. The difference between production and export is domestic consumption.

Source: Consumption estimate from S. A. R. Kadhim and A. Al-Janabi, "Domestic Energy Requirements of OPEC Member Countries" (Paper presented at the OPEC seminar OPEC and Future Energy Markets, October 1979). Compiled by author.

TABLE 4.3

Current Reserves and Production and Likely Exports in the 1980s (millions b/d)

	Proven Reserves[a] (billion barrels)		Reserve to Production Ratio	Production/Domestic Consumption/Exports 1979[b]			Production/Domestic Consumption/Exports 1985[c]			Production/Domestic Consumption/Exports 1990[c]		
	1973	1979	1979									
Saudi Arabia[d]	138.0	165.7	49	9.20	0.25	8.95	6.3	0.46	5.84	6.3	0.84	5.46
Iran[e]	65.0	59.0	54	2.98	0.64	2.34	3.0	1.04	1.96	2.7	1.66	1.04
Iraq[f]	29.0	32.1	26	3.38	0.20	3.18	4.0	0.34	3.66	4.5	0.53	3.97
Kuwait[g]	64.9	66.2	81	2.25	0.04	2.21	1.5	0.06	1.44	1.0	0.09	0.91
Qatar	7.0	4.0	22	0.50	0.01	0.49	0.5	0.01	0.49	0.5	0.02	0.48
United Arab Emirates[h]	22.8	31.3	47	1.82	0.05	1.77	1.8	0.07	1.73	1.0	0.11	0.89
Neutral Zone	16.0	6.5	32	0.56	—	0.56	0.5	—	0.50	0.5	—	0.50
Total: Persian Gulf	342.7	364.8	48	20.69	1.19	19.50	17.6	1.98	15.60	16.5	3.25	13.25
Venezuela	13.7	18.0	21	2.34	0.29	2.05	1.9	0.50	1.40	1.3	0.79	0.51
Nigeria[i]	15.0	18.2	21	2.33	0.16	2.17	1.9	0.27	1.63	1.8	0.42	1.38
Libya	30.4	24.3	32	2.06	0.08	1.98	1.6	0.17	1.43	1.2	0.31	0.89
Indonesia[i]	10.0	10.2	17	1.60	0.36	1.24	1.6	0.59	1.01	1.5	0.92	0.58
Algeria	47.0	6.3	15	1.14	0.14	1.00	0.9	0.27	0.63	0.9	0.45	0.45
Gabon	1.1	2.0	26	0.21	0.02	0.19	0.2	0.03	0.17	0.2	0.05	0.15
Ecuador	5.7	1.2	16	0.21	0.06	0.15	0.2	0.09	0.11	0.2	0.12	0.08
Total: OPEC	465.6	445	40	30.58	2.30	28.28	25.9	3.90	22.00	23.6	6.31	17.29
OPEC as % of world[j]	70	69	—	50	—	83	—	—	—	—	—	—

[a] Proven reserves denote reserves that can be recovered with current prices and technologies. Excludes secondary and tertiary recoveries. Secondary recovery alone can boost reserves in the Middle East by at least 50 percent. The proven reserve figures widely quoted in the literature are questionable to say the least. Often fields of 50 million to 500 million barrels that have a cost of production of over $1/b are considered too small, too expensive, and are not included in reserve figures. The figures for Iraq and Kuwait are generally believed to be wrong. Iraqi reserves are believed to be 4 to 5 times larger than the quoted figure—second only to Saudi Arabia, while Kuwaiti figures are thought to be exaggerated. Most OPEC nations as a matter of policy do not declare their reserve figures and the information is generally provided by the oil companies.

[b] Reflects January–October 1979 production.

[c] Production estimates are based on the author's expectations of production cutbacks due to political/economic and technical reasons. These estimates exclude the possibility of major political upheavals. Domestic consumption forecasts are quoted from the OPEC Secretariat.

[d] Many powerful voices in the Saudi royal family prefer a cutback in production to 4 million to 4.5 million b/d. Nevertheless, barring major political changes, one can assume that the Saudi production preference in the 1980s will be at a level with which its massive gas-gathering project for associated gas can operate at full capacity. The gas-gathering project, which is estimated to cost between $14 billion and $21 billion, has a collection target of 3.3 billion cubic feet a day and will be ready by 1982. To feed the new system, an oil production level of 5.7 million b/d from Ghawar and Abqaiq fields together with 550,000 b/d from Berri field are sufficient. Therefore, a production level of 5.3 million b/d is assumed here to be the likely production for the 1980s. It is highly unlikely that the Saudis will invest heavily in expansion of capacity as a larger capacity will place them under international pressure to increase production.

[e] Iran's production is expected to remain relatively steady during the 1980s, although an increasing share will go to domestic consumption. It is assumed that by 1990, Iran will try to export one million b/d to pay for its import needs. Technical problems by this time will seriously curtail additional production.

[f] The poor relationship between the postrevolutionary governments of Iraq and the oil companies in the 1960s was the reason behind ignoring the development of Iraqi oil reserves and capacity. Indeed, the expansion in production of a number of other Middle East producers came at the expense of Iraq. The situation was reversed in the 1970s when Iraq, nationalizing its oil, heavily expanded capacity and production. In 1975, when OPEC production as a whole fell by 11 percent due to declining demand, Iraqi exports rose by 15 percent. Despite temporary cutbacks occasionally, it is clear that the Ba'athist government of Iraq in its quest for influence and power in the region will push production and capacity higher in the 1980s. This means that

TABLE 4.3 (cont.)

Iraq will probably be the only country in OPEC expanding production. Iraqi capacity in 1990 is expected to be around 5 million to 6 million b/d.

gKuwait has expressed its intention to cutback production to 1.5 million b/d in the short term and maintain a 100:1 reserve/production ratio in the long term. Like Saudi Arabia, Kuwait's flexibility is restricted to some extent by its associated gas facilities. The difference with Saudi Arabia is that all the gas is used domestically: 33 percent for desalination; 25 percent for oil company use; and 9 percent for reinjection. Together with LPG plans, ideally 3 million b/d production is required for full capacity utilization, a production level long abandoned by Kuwait. Still, production is expected to decline, but only gradually. Like Saudi Arabia, UAE, and Libya, accumulated surplus investments abroad will bring in large interest income in the 1980s.

hIncreasing tensions between oil-producing and non-oil-producing emirates, together with uneven distribution of oil income within the seven sheikdoms that form UAE, will not allow a sharp cutback in production in the next five years. If the federation is to continue in one piece, oil income will have to be channeled to poorer sheikdoms in larger quantities. Still, by 1990 the lower production level of 1 million b/d is thought to generate sufficient income for UAE's needs.

iAs the two most populous nations of OPEC, these countries' need for ever increasing oil income is expected to remain high. Indonesia, the poorest OPEC country, is expected to maintain production at full capacity, while Nigeria is expected to reduce production by only 215,000 b/d compared to its declared target of 2.15 million b/d for 1980. Recent reports indicate an expansion in the production capacity of both nations is likely in the 1980s. According to the Oil Daily (November 1, 1979), Indonesia's exploration and development efforts have increased substantially and a production capacity of 2 million b/d by 1985 is likely. During 1990–2000 a capacity of 2.5 million b/d is not impossible. For Nigeria, the recent report by the U.S. Department of Energy and U.S. Geological Survey Report on the Petroleum Resources of the Federal Republic of Nigeria (October 1979) indicates that with increased drilling, prospects for capacity expansion to 3.5 million b/d is there. The report, however, anticipates a production range of 2 million to 2.5 million b/d could be maintained during the latter part of the 1980s through the year 2000. In the case of both countries the available production capability will not necessarily mean that they will produce at those levels.

jData not available (—).

Source: Compiled by author.

persist beyond 1981. It was created by declining petroleum consumption in the industrial countries (as a result of the world recession), conservation, and overproduction of petroleum by Saudi Arabia (1 million b/d above their announced ceiling of 8.5 million b/d). This oil glut, like the one in 1975, diverted attention from the fundamental problem of petroleum supply. This time, however, the glut promises to be short-lived. Saudi Arabia will probably reduce output by 1 million b/d in 1981 and the world economy will probably move out of the recession, producing a balanced market by mid-1981.

Over the next 10 to 15 years, the outlook seems very grim. Even without major political upheavals, there is likely to be large reductions in exportable crude, both through production cutbacks and through large increases in domestic petroleum consumption.

Table 4.2 presents two scenarios for the late 1980s and 1990s: constant production at the 1980 level and declining production. The possibility of production higher than 1980 levels, except for temporary periods, is out of the question and

Table 4.2 shows that even if production does not decline, the mere increase in domestic demand for petroleum will reduce exports by 6 percent by 1985 and 16 percent by 1990. If restrictive production policies are adopted, exports could well decline one-third by 1990. For the year 2000, domestic consumption of 16.7 million b/d within OPEC could reduce exportable supplies to only 11.3 million b/d—even with the present rate of production.

Table 4.3 shows production and export possibilities for individual OPEC countries. This table takes into account plans, options, perceptions, and technical/resource problems that OPEC nations face. This tabulation provides a realistic outlook for future OPEC production and exports.

EXPORT POLICY

Though it is not possible to define a unified export policy for OPEC in the future, producers will probably follow certain practices. First, they will further reduce sales made on a preferential basis to the large international oil companies (e.g., "equity" or "buy-back" oil). In 1973, 92 percent or 27.9 million b/d of the international oil companies' petroleum supplies from OPEC were sold on some sort of preferential basis. By 1979, this ratio had fallen to 58 percent or 17.5 million b/d. At the same time, direct state-to-state deals rose from 1.5 million to 5 million b/d, while direct commercial sales rose from 1.7 million to 7.8 million b/d. In total 42 percent of OPEC exports were direct sales. The change in traditional sources of supply has meant that the international oil companies cannot fully meet commitments to third parties. Between 1973 and 1979, sales to third parties were slashed by 50 percent, from 6.8 million to 3.4 million b/d.[8] These trends are likely to continue in the future. By 1990, OPEC may well be handling three-fourths of its exports directly. Third-party purchasers will not be able to depend on the international oil companies for supplies. State-to-state deals will be encouraged by the oil exporters. Preference will be given to national oil companies, particularly if the buyer is from the developing world. Those who need oil will have to deal directly with the oil-exporting nations.

Second, spot sales are likely in periods of acute shortage. As OPEC nations directly control more of their exports and more sales are made directly to other nations, spot sales will disappear.

Third, long-term supplies are not likely to be awarded. Contracts lasting 6 to 12 months are expected to become the norm.

Fourth, when the oil is "lifted" (i.e., picked up) producers will impose more restrictions on where oil may be shipped. The international oil companies will not be able to reroute supplies during embargos, which will hurt the embargoed nations severely.

Fifth, oil sales will be in package deals. The packages may include the following:[9]

- Investments in exploration or establishment of petrochemical, refining, or other industries.
- Acceptance of refined products and petrochemicals even if there is a surplus of these products in the consuming countries.
- Transport by OPEC-owned tankers.
- Arms sales or technology transfers.
- Major concessions from the industrial world toward the poor nations.
- Indexation of OPEC investments in the industrial world.

When OPEC nations decide who will receive crude oil in the 1980s and 1990s, purely commercial considerations will be outweighed by other factors. Consumers will have to accept a host of political and semieconomic conditions. In the view of the OPEC nations, they are exhausting an asset that took millions of years to be created. It is perhaps their only chance to accomplish their goals. Differences in political ideology or alliances are not going to change radically their course of action.

OPEC PRICE HISTORY

OPEC price policy has gone through a number of phases. From 1960 to 1971, nominal prices were frozen to shield the oil producers from the continuous trend of declining prices. There were declines in the real price of crude because of inflation, and spot prices remained below official prices. Because of inflation, from 1971 to October 1973, OPEC prices were increased through negotiations with the oil companies. Thereafter, nominal prices were increased unilaterally by the OPEC Ministerial Conferences, though prices fell in real terms until the third quarter of 1979. The Saudi Arabian "marker" crude of 34° API was used as a yardstick; other crudes were priced according to differences in quality and distance from consuming areas. Spot prices occasionally rose above official prices for short periods of time.

The Iranian revolution of February 1979 sent the oil market into disarray. Spot prices rose substantially above official contract prices, reaching $45 per barrel; more than double the $18 per barrel price of marker crude. Though spot sales represent only a small portion of the total oil trade, they are significant because they are a signal to the oil exporters that official prices can be raised. It is important to note that 1979 price increases were market-induced rather than administratively decided by OPEC.

The significance of the 1979 price increases goes far beyond the immediate impact of higher prices. In 1979, more oil was produced and exported than in 1978 despite the Iranian supply cutback. Why did prices rise so sharply? There were several factors.

First, the conventional oil distribution mechanisms broke down. Traditionally, the international oil companies received the bulk of oil exports and distributed them to those who needed supplies—their own affiliates and third parties. The Iranian situation denied the large oil companies their usual access to crude. Other OPEC nations also began to sell their oil directly to state oil companies and smaller independents. For instance, Japan was denied 1 million b/d of oil by the majors. Since there was no mechanism to fill the vacuum, the Japanese went into the spot market, pushing up spot prices. In short, the number of actors in the market increased substantially. While there were sufficient volumes of oil available, those who needed it could not get access to it without using the volatile spot market.

Second, the Iranian crisis signaled a period of supply instability in the Persian Gulf, the most important oil region of the world. The fear of more interruptions from this region, whether justified or not, has pushed up stockpiling to a record volume: 4 billion barrels. Unlike the past, such stocks will probably not be drawn down substantially throughout the 1980s because private oil companies as well as governments believe supplies are insecure.

Third, the fall of the former Iranian government substantially reduced Saudi Arabia's flexibility. Iran was the second largest exporter of crude and, since 1975, a natural, if unwilling, ally of the

Saudis is pushing for moderate policies on production and prices.
Now, Saudi Arabia is left alone to argue for moderation, with occasional help from the UAE. In 1979, the Saudis kept their price at $18
per barrel—$4 below similar quality crude—in order to keep these
other prices from rising sharply. They failed, and later had to raise
their prices to $24, $26, $28, to $30 per barrel in September 1980
and over $33 and less than $35 in 1981. Though they export more than
one-third of OPEC's oil, their former position as the "swing producer"
has been seriously eroded. By stretching their output to capacity, they
have left themselves little room for maneuvering. Though they may
succeed in temporarily stalling the rise in prices, in the medium and
long term they are not likely to succeeed. Soon they will find this
battle not worth fighting. They will bow to domestic and OPEC pressure to reduce supplies, eliminating the current mini-glut they created.
Although the importing industrial nations contributed to the glut by
reducing their consumption of petroleum, self-congratulations are not
in order, since the Saudis have been producing far above their own
needs. A 1 million to 2 million b/d drop in Saudi production, which
could happen overnight, would change the picture totally.

With the decline in Saudi influence on the world oil market, the
Saudi Arabian marker crude has lost its credibility as the yardstick
for determination of other oil prices. Instead there is a "theoretical
marker crude" at $32 per barrel, which is used in many transactions.

DIFFERENTIALS FOR LIGHT, LOW-SULFUR CRUDE

The traditional issue of "differentials" used to price crudes of
different quality in comparison to the Saudi marker crudes is now
viewed in a totally different light. Neither the Saudi marker nor the
theoretical marker is consistent with prices of $37 per barrel now
charged by Algeria, Libya, and Nigeria. Under the old ground rules,
the differential would have been $2 to $3 per barrel. Those analysts
who expect the current differentials to return to "normal" may have
to wait forever.

To understand the issue of differentials, one must look at OPEC
exports in terms of regions from which the exports originate. The
behavior of individual nations is more dependent on the market within
the region that on markets in other regions. The African OPEC members produce light, low sulfur crudes, which command especially
high prices in Europe and particularly the United States. These consumers have gradually shifted toward these crudes because they cause
fewer environmental problems and produce more gasoline. The growing demand for these light crudes has partially segregated this market
from the market for other OPEC exports. It just happens that two of

the most radical OPEC states, Libya and Algeria, are in the African region.

Consumers of light crude are prone to a higher degree of instability. A small interruption from the Mediterranean could have a more damaging impact than a larger interruption elsewhere. Such dependencies, of course, are not irreversible, but it could take substantial refining investments, the relaxation of environmental restrictions, and a lead time of two to three years to reverse the trend.

OIL PRICING FORMULAS

Since 1974 the OPEC nations have considered a number of bases for uniform oil pricing, including: alternative energy costs; the price index for OPEC imports; the price index for exports from the industrial nations; spot prices; expressing prices in Special Drawing Rights (SDRs), or another devaluation-proof "basket," rather than in U.S. dollars; and a method recommended by OPEC's Long-Term Strategy Commission (LTSC).

One of the earliest and perhaps the most powerful of the arguments for raising oil prices was based on the idea of parity between the price of oil and the cost of alternative sources. This argument was first given worldwide attention in 1973 when OPEC raised the government take to $7 per barrel, OPEC's estimate of the average cost of energy from alternative sources. This argument is persuasive, not only because it has an economic rationale behind it, but also because it is appealing to some academics in the West. One, of course, has to define "alternative energy costs." Are we talking about primary products or end use? Do we take into consideration costs associated with environmental protection? Are we considering synthetic fuels alone? Are the cost levels associated with small production or laboratory testing reliable enough to compare them to oil without utilization of economies of scale? As OPEC prices rose, so did the cost of alternatives—but not all alternatives. Clearly, at current prices, nuclear power, solar water heating, and use of coal for power generation are competitive with oil. Oil production from shale and tar sands is nearly competitive. The average cost of synthetic fuels is still considered by the OPEC Secretariat to the $50 to $60 per barrel of oil equivalent, and is higher than crude. However, without large-scale operations, no one will know for sure whether the costs of alternatives are competitive with present crude prices. Alternatives surely will be competitive in the 1980s, after the expected increase in the price of crude.

OPEC does not attempt to price its oil to keep alternative sources out of the market. The argument for parity is a good one and may be

used to justify higher oil prices, but OPEC decision making is simply
not based on this type of analysis. Decisions on prices apply only for
six months and are reached by a struggle between the various power
centers in the organization. Indeed, because they expect a shortfall
in petroleum supplies, and experience major internal political pressures to reduce production, many OPEC nations would welcome the
development of alternative energy sources.

Sentiment has shifted between pricing by an index and pricing
according to the spot market, depending on the status of the market.
If the market is soft, OPEC emphasizes the need for indexation; if
the market is tight, the free market is favored. This is a good strategy for the OPEC nations, since it tends to keep prices high.

Indexation would end infighting over price increases and make
it difficult to break ranks. This method also would eliminate the
Ministerial Conferences that are now held in June and December of
each year. These meetings simply attract much publicity and give
importers the opportunity to exert pressure to hold price increases
to a minimum. However, the OPEC nations clearly realize that indexation is useless if the market is tight. First, the OPEC nations have
no incentive to stick to indexation. They might even consider leaving
the organization if they were bound to sell oil at lower prices than
they could command in the market. Second, indexation can work only
for government-to-government sales. As long as there is a private
market, indexation has limited applicability. Even if producers stick
to indexation during a crude shortage, prices will rise through resales
by the oil companies and the consumer will pay the market price. If
someone is to reap the benefits of high prices, why shouldn't it be the
OPEC nations themselves?

Two general types of indexes have been proposed within OPEC.
The first would reflect the increase in the cost of OPEC imports; the
second would reflect the price of exports from the industrial nations.
The difference is important. During the 1970s, the prices of OPEC
imports consistently increased between 25 to 35 percent annually,
while the price of industrial exports has increased only 5 to 15 percent annually. The gap was large enough to make some OPEC nations
think that they had been deliberately discriminated against by the industrial nations. A closer examination of the components of OPEC
imports has shown that OPEC customers did pay more than other customers, but that such discrimination was not necessarily by design.
Part of the difference can be explained by the type of goods OPEC
nations purchased: super-sophisticated weaponry, advanced petrochemical plants, nuclear power plants, and so forth. In many cases,
only a few companies produce such advanced technology, so there is
little or no competition. The OPEC nations had excess funds, did not
look far for the best deals and the best options, and so paid higher

prices. Had these types of purchases been eliminated from the OPEC import index, the increases in that index would have been much smaller. Another part of the difference can be explained by the difference between f.o.b. (free on board) and c.i.f. (cost, insurance, freight) prices. Often the OPEC nations paid premiums for quick delivery of items they ordered, or demurrage fees to ships waiting in crowded terminals.

Despite all the arguments for indexation, OPEC has not yet applied such a system and probably will not successfully. The export price index is so strongly opposed by Algeria, Libya, Iraq, and Iran that it has virtually no chance. OPEC may try to use an import index, but it will not work because it requires price hikes that are too high for a soft market. In summation, indexes are doomed because they raise prices either too much or too little without consideration of the spot market, and because they require an unlikely degree of discipline within OPEC.

As a supplement or alternative to indexation, the OPEC nations have sought to insulate themselves from the persistent decline in the exchange value of the U.S. dollar. Since oil is priced and oil payments are received in dollars, the OPEC nations have lost a great deal of money as the dollar has declined in value. To protect themselves the OPEC nations initially considered using International Monetary Fund Special Drawing Rights (SDRs), which are based on a "basket" of 16 currencies, as the unit of account. Later, OPEC realized that use of SDRs would not shield them sufficiently against dollar fluctuations, since the dollar accounts for 33 percent of the SDR's value. OPEC has also considered using two other baskets of 9 to 11 currencies, including the dollar, weighted according to OPEC's own import pattern. Such baskets, referred to as Geneva I and Geneva II, were temporarily used in the early 1970s to measure losses due to dollar devaluations. Other suggestions, such as using deutsche marks, Swiss francs, or Japanese yen, were also considered. However, the Europeans and Japanese are unwilling to have their currencies singled out as the medium of exchange, since peaks in currency demand due to monthly oil payments could play havoc with their exchange rates. No one currency except the U.S. dollar could accommodate such large transactions, which amount to 15 to 20 percent of total world trade. So far, OPEC has not decided whether to use a basket of currencies to express prices.[10] However, if and when such a system is adopted—no matter what denomination the price is in—the final payment must be in dollars. That is, payments will be converted to dollars at the current rate of exchange. This is not likely to have a long-term impact on the dollar exchange rate, except for the initial one-time impact of the vote of no confidence in the U.S. currency by the OPEC nations.

The latest formula for setting prices in the one that came out of

TABLE 4.4

Actual and Hypothetical Price Movements under LTSC Formula

| | LTSC Indexes | | | | | | | | Price ($/b) | |
| | Inflation | | Exchange Rate | | GNP | | Aggregate | | | |
	Index	% Growth P.A.c	Index	% Growth P.A.	Index	% Growth P.A.	Index	% Growth P.A.	Under LTSCa	Actualb
1973	100	–	100	–	100	–	100	–	–	5.04
1974	116.8	16.8	99.7	-0.3	100.3	0.3	116.8	16.8	10.84	11.25
1975	134.4	15.1	102.3	2.6	99.3	-1.0	136.5	16.9	12.66	12.38
1976	140.5	4.5	97.4	-4.8	104.5	5.2	143.0	4.8	14.80	11.51
1977	152.4	8.5	99.6	2.3	103.3	3.7	164.4	15.0	15.50	12.70
1978	158.0	3.7	110.2	10.6	112.3	3.7	195.5	18.9	17.82	13.33
1979	168.8	6.8	114.1	3.5	116.2	3.5	223.8	14.5	21.19	18.00
1980d	–	–	–	–	–	–	–	–	24.26	24.00

aPrice increases are compounded by previous year changes.

bArabian marker crude, price in the first half of each year. Figures for 1973-76 are posted prices and for 1976-80 official sales prices.

cPer annum.

dData not available.

Note: OPEC Long-Term Strategy Committee (LTSC) formula, March 1980.

Source: Petroleum Intelligence Weekly, May 12, 1980.

the OPEC Long-Term Strategy Committee (LTSC) in March 1980.
This formula—the architect of which is Sheikh Yamani, the Saudi oil
minister—proposes three components for the price of oil:

- An antiinflation index weighted on exports of manufactured
  goods from the member countries of the Organization for
  Economic Cooperation and Development (OECD), and consumer cost of living in the OECD countries.
- An automatic exchange rate adjustment (to safeguard against
  dollar weakness) based on a combination of the U.S. dollar
  and the nine-currency Geneva I basket.
- An increase in the price of oil in line with the real rate of
  growth in the GNP in OECD countries.

The first two components are meant to keep the real price of
oil from falling while the third component would raise the real price
of oil gradually to reach the cost of alternative energy. The price
arrived at would be a minimum or "floor" price. If market shortages
were acute, the floor price would be raised to reach market levels
and the three components activated from the new base. Under this
regime, price would be adjusted on a quarterly basis and the importing countries could forecast their oil receipts in advance with some
accuracy.

Table 4.4 shows how price changes under this regime would
have compared with actual changes since 1973. Prices under this
formula would have been similar to actual prices, except for 1980
when Arabian marker crude rose from $24 per barrel in January 1980
to $30 per barrel in September 1980. In all years except for 1976,
nominal oil prices would have risen by at least 15 percent.

In the future, the LTSC price formula would cause oil prices to
rise by around 5 percent annually in real terms, assuming real GNP
growth of around 2.5 to 3 percent and expected dollar devaluations
against a basket of currencies of 2 to 2.5 percent. In nominal terms,
the annual price increase for 1980-85 would be around 10 to 15 percent.

The LTSC formula was approved by ten OPEC nations in May
1980, with Iran, Libya, and Algeria expressing reservations. The
final decision on this formula is expected to be made in November
1980 during the summit of OPEC heads-of-state in Baghdad. The
difficulty with adopting this formula is that an initial base must be
set. But with variations of $7 per barrel between Saudi oil prices and
Libyan prices (or $5 per barrel between Saudi Arabia and Iran), such
a policy cannot be implemented. The LTSC formula seems doomed to
failure because it suffers from the same weakness of other indexes—
failing to take account of market conditions.

PRICE PROJECTIONS

The implication of the events in 1979 is that the real price of oil will not be allowed to decline again. To understand the likely trend of oil prices, one should try to understand the process of decision making within OPEC. Oil price modeling is misleading and invariably leads to false expectations if it is based on demand and supply considerations or on hypothetical OPEC revenue needs based on absorptive capacity (which can never be statistically substantiated, even if theoretically sound). OPEC's actual decision-making process is far less complicated. Decisions are arrived at by simple consensus or, failing that, on an individual basis. The upper ceiling to OPEC's oil prices is set by the international security considerations of one or all member nations. Each nation tries to avoid price hikes that would invite wars or cause catastrophic changes endangering that nation's own security. Close ties with consumer nations may induce some exporters to be more moderate than others, although as we have seen, the market can accommodate even a $10 per barrel difference between similar quality crudes. As long as there is the likelihood of a shortage, there is no need for unity in prices. Indeed, it is economically irrational for OPEC nations to seek unified prices. To deduce from the lack of price unity that OPEC is in danger is wishful thinking.

Oil prices in the 1980s are, thus, far less dependent on demand than supply. If the consumer nations manage to save a few million barrels per day in crude imports or, on the other hand, increase their imports by a few million barrels per day, the effect on prices will not be significant. In the former case price will be increased by regular increments; in the latter, by large jumps. Indeed, the producer's own national security considerations are more important than demand. As the oil market tightens, spot prices will lead the way. Official oil prices will be increased to near spot levels and maintained through production cutbacks. If the new levels endanger their national security, OPEC nations will proceed with caution. They will gauge how far they can go by considering developments in the industrialized importing nations. Once official prices are raised, if demand declines, prices will be administratively maintained and exports will fall. One year's lower production will become the next year's ceiling, ensuring a persistent shortage in the market. Lower production and exports will, in turn, reduce the domestic political pressure on the oil exporters.

Oil prices are likely to rise between 5 and 10 percent annually in real terms. The view generally held by the industrial nations and the World Bank is that oil prices will rise by 3 percent annually during the 1980s. The fact that the LTSC formula (which in effect advocates a 5 percent annual real price increase) was not accepted

TABLE 4.5

Oil Price Scenarios in the 1980s
($/b)

	1980	1981	1985			1990		
			Low	Medium	High	Low	Medium	High
Average OPEC prices (constant 1980 prices)	31	31	31	38	45	31	48	73
Average OPEC prices (current prices)[a]	31	33	43	52	62	61	92	136

[a]Assumes 7% annual rate of inflation.

Note: Low indicates no real price increase, medium 5% real price increase, and high 10% real price increase per year.

Source: Compiled by author.

by some members reflects their expectations that oil prices will rise by more than 5 percent annually.

Table 4.5 presents three scenarios of oil prices. In the 1980s oil prices are likely to increase 5 to 10 percent annually. By 1990, we can expect real prices of $48 to $73 per barrel. These price estimates do not take into account the possibility of major political upheavals or revolutions in the key OPEC nations.

THE FUTURE OF OPEC

OPEC has lost much of its power and usefulness as a price-setting organization. It provides a way to raise or maintain prices in times of glut or balance in the oil market. But once shortages become persistent, OPEC nations will go their own way in deciding prices and production levels. Because the market will probably become tight, we can expect different producers to charge different prices. OPEC members will individually find the maximum price each can charge through trial and error, making monthly or quarterly adjustments.

At the same time, the Iranian revolution has caused a redistribution of power within the organization in favor of the smaller producers. Smaller producers can, individually or in a group, upset the world oil market by temporary or permanent production cutbacks. Indeed, even the threat to do so could have a major effect in a shortage situation.

Still, it is unlikely that OPEC will vanish. OPEC might perform a great number of functions besides price setting. Many of these functions have been discussed in the LTSC report in detail and will not be elaborated here. Briefly, OPEC can become a forum for representing members' interests in a range of areas, from relations with industrial countries to economic assistance policies toward the developing world. OPEC can also serve as a trading bloc that could obtain concessions for itself and other LDCs.

OPEC faces a number of specific challenges. First, the OPEC Gas Committee has had little success in the coordination of policies toward pricing and exports of natural gas. Similarly, the OPEC Petrochemicals Committee has had little success: petrochemical plants are mushrooming everywhere in the Persian Gulf even though the world petrochemical industry is suffering from surplus capacity and some industrial nations are preparing to impose tariffs and quotas on OPEC petrochemical exports. In both cases, the primary reason for the lack of success has been Saudi Arabia's opposition to a unified approach. In the 1980s, as Saudi influence weakens, more coordination is expected on these fronts.

Third, though the possibilities of downstream activities abroad are extremely dim, OPEC nations are expected to increase massively their refining capacity and their oil tanker fleet. A survey of the orders[11] already placed and the plans of the OPEC exporters indicates that by 1990 30 to 50 percent of OPEC exports may be in refined products and at least 20 to 25 percent will move in OPEC tankers. These activities will increase the effect of embargoes and the political influence of the OPEC nations.

The most immediate problem facing OPEC is the plight of the developing nations. Though OPEC's surplus has swelled from $1 billion in 1978 to over $100 billion in 1980, OPEC's economic assistance program (which is far more generous than that of OECD's Development Assistance Committee) has not kept pace with the more than doubling of the LDCs' oil import bills. At the same time, the commercial banks, which have become the major supplier of funds to LDCs, are hesitant to recycle petrodollars to the needy nations, fearing defaults and consequent shocks to the world banking system. The banks are urgently seeking insurance mechanisms from OPEC, OECD, and the International Monetary Fund. OPEC is under pressure to act quickly, from domestic political forces and in international forums in which LDCs are openly beginning to voice their criticisms. The old assistance mechanisms are not going to be sufficient.[12]

NOTES

1. The actual government's take was $.80 per barrel in 1970 and $7 per barrel in 1975. The remainder was either production costs, oil company profits, or "national" price that was often discounted.

2. Petroleum Intelligence Weekly, New York, November 19, 1979.

3. Morgan Guaranty Trust Company, World Financial Markets, New York, November 1979.

4. World Bank, World Development Report (Washington, D.C., 1979).

5. See F. Fesharaki, Revolution and Energy Policy in Iran (London: The Economist Intelligence Unit, 1980).

6. World Financial Markets, p. 4.

7. "Back-to-back" deals are those that involve disguised discounts, e.g., an oil producer sells oil at the official price but secretly agrees to buy other goods at an inflated price.

8. Petroleum Intelligence Weekly, February 25, 1980.

9. F. Fesharaki, "The Evolution of Petroleum Contracts" (Paper presented to the Workshop on Mineral Policies to Achieve Development Objectives, East-West Center, Honolulu, June 9-11, 1980). Available from author. For details of OPEC activities in refining, petrochemicals, and tanker transport, see F. Fesharaki, An Analysis of Petroleum Section Development Plans and Perception of the Major Oil Exporters (New York: United Nations, Energy Division, 1980).

10. In the 1974 OPEC Ministerial Conference in Quito, the use of SDRs was officially endorsed, but in practice no country adopted it.

11. Fesharaki, An Analysis of Petroleum Section Development Plans.

12. For a discussion of new options open to OPEC see F. Fesharaki, "OPEC vs. LDC's Energy Needs," Paper presented at the Workshop on Energy and the Developing Nations, Electric Power Research Institute and Stanford University, 1980.

5

WORLD OIL OUTLOOK

Tom E. Burns

All too often, those in the oil business wind up talking to ourselves most of the time and, of course, it is very easy to convince people of like mind. It is essential to seek out different views and data, and it should be obvious that a social scientist looks at OPEC with entirely different eyes than a reservoir engineer.

It is also important, however, to remember that in all things fashion plays a role. Yesterday's heresy becomes today's common knowledge or conventional wisdom, and there should be caution about becoming too pessimistic, just as there should have been caution some years ago about being much too optimistic about oil supply and the world's energy situation.

One should get a little bit nervous when forecasts begin to look the same. Once that begins to happen, observers should become wary and begin to wonder what it is that all of us collectively are missing. If there is no disagreement, we all may have missed the critical points.

The oil forecast of the Standard Oil Company of California falls somewhere in the mid-range. The CIA (see Chapter 3) presented a somewhat more pessimistic view of world supply, while people like Herman Kahn have stated about the opposite extreme.

Standard Oil's basic assumptions are critical, but hardly surprising: moderate economic growth, moderate energy growth, world stability, and real energy price increases. There is no attempt to forecast discontinuities, although they undoubtedly will occur. When future analysts review the history of the 1980s and 1990s, they will undoubtedly see a number of aberrations, like the one in 1979 in Iran that triggered the shortages thereafter. Although discontinuities like this have a cumulative effect on future trends, each one may be hardly perceptible on the scale of world energy consumption. The disruption in supply of 1974-75 was definitely perceptible. The same was true for 1979 but there could well be similar events happening in the future that are going to be just as influential, but perhaps not as readily apparent as the first two oil shocks.

Standard Oil forecasts that world energy consumption is going to continue to grow through the end of the century but at significantly slower rates than it has historically (see Figure 5.1). These lower growth rates will come about as the result of higher energy costs, which will lead to continued improvements in the efficiency of energy use as well as slower economic growth. Energy consumption in the industrialized areas of the world will grow more slowly than in developing areas. U.S. energy consumption will still increase by over one-third from a present level of about 40 million barrels a day oil equivalent (boe) to 55 million boe by the end of the century. Western Europe's energy growth will be intermediate to Japan's, where energy consumption is still expected to double in the next 20 years. The Sino-Soviet area will continue to show rapid growth based upon its resources, level of development, and population pressures. This is particularly true, of course, in China.

Future energy requirements will be supplied by a gradually changing mix of sources (see Figure 5.2). Conventional oil production will continue to grow through 1990 and gas through the end of the century. However, their combined share of total energy supply will decline from 63 percent in 1980 to 50 percent in the year 2000. Looking at the various sectors we see that growth rates will be lower for oil, gas, and hydropower in the years to come. Increased energy supplies must, therefore, come predominantly from coal and nuclear as well as from more exotic sources such as solar, geothermal, and biomass. Standard Oil's nuclear forecast is based upon plants that are either operating or under construction plus planned installations in nations that are still contemplating a nuclear future such as France, Japan, and Brazil. It is obvious that after Three Mile Island, the nuclear future is still very cloudy at best.

The scale of Figure 5.2 shows the problems facing the solar industry. Even at growth rates that call for the contribution from solar and miscellaneous sources to double every eight years, they still only provide about 5 million b/d in the year 2000.

Let us turn for a moment to Standard Oil's forecast of world oil supply (see Figure 5.3). Conventional crude production should peak in the 1990s and begin to turn down. However, continued growth in natural gas liquids (NGL) and synthetic oils will permit total oil supply to grow by just over 1 percent per year to the year 2000. Oil production in the Sino-Soviet area will continue to grow, but it will not keep pace with consumption.

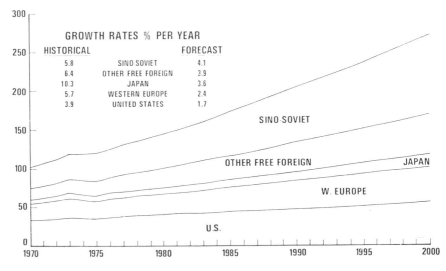

FIGURE 5.1

World Energy Consumption
(million b/d)

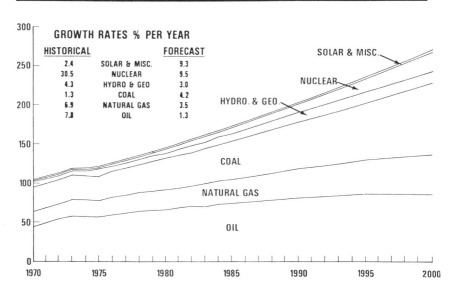

FIGURE 5.2

World Energy Supply
(million b/d)

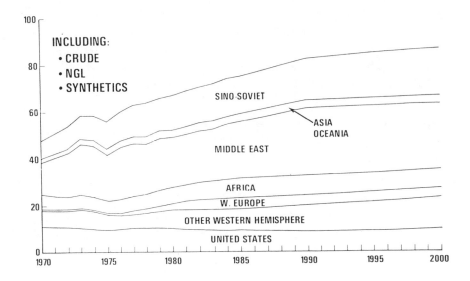

FIGURE 5.3

World Oil Production
(million b/d)

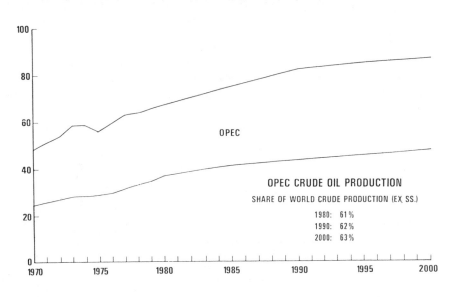

FIGURE 5.4

OPEC Oil Production
(million b/d)

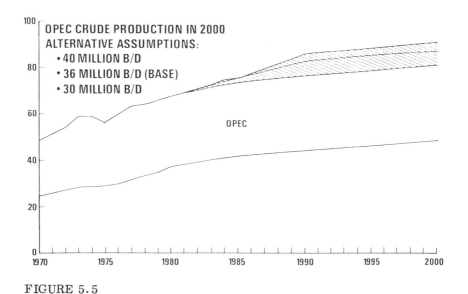

FIGURE 5.5

Forecasts of OPEC Oil Production
(million b/d)

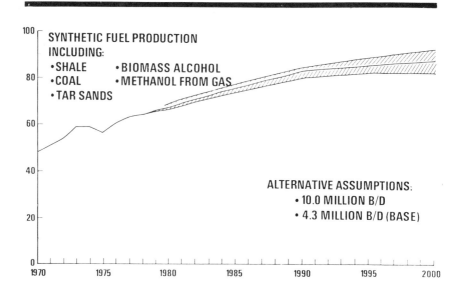

FIGURE 5.6

Synthetic Fuel Production
(million b/d)

This represents a change from Standard Oil's previous forecasts and is more in line with CIA forecasts for the Sino-Soviet area. Standard Oil now expects that net imports from the Sino-Soviet area to the West will taper off in this decade, be negligible in the 1990s, and could well become net importers during that period.

U.S. oil production will continue to decline until 1990, after which synfuel growth will be able to slightly more than offset further declines in the production of conventional oil.

In view of the many uncertainties ahead, it is essential to examine some of the critical forecast assumptions in more detail. Of course, one of the most critical is the role of OPEC. Standard Oil's forecast of potential OPEC production is based on published and announced reserve data, production limits, and investment programs. OPEC could produce crude oil at a peak rate of about 36 million b/d during much of the 1990s (see Figure 5.4). Total reserves could, on a theoretical basis, support even higher production rates during this period. OPEC's relatively high reserve to production ratios means that natural declines will set in later in the OPEC nations. Therefore, OPEC's share of world crude production, excluding the Sino-Soviet area, will actually rise during this period. However, since continued growth in natural gas liquids and synthetics over the same period is anticipated, OPEC's share of total oil supply will decline slightly.

Now just how good is this forecast of OPEC crude production? It was not long ago that most observers, Standard Oil included, were predicting OPEC production rates of 40 million b/d in the 1990s. Certainly the reserves could theoretically support such a rate. On the other hand, EXXON, Walter Levy, and many others now suggest that OPEC crude production will level out at the current rate of about 30 million b/d. The CIA and others are even more pessimistic, believing that OPEC crude production will begin to decline from the current level.

Standard Oil's forecast falls somewhere between these extremes (see Figure 5.5) and while that is by no means a test of a good forecast, it is comforting to know that the OPEC Long-range Committee also predicts that production of OPEC crude could be about 36 million b/d in the 1990s.

Let us look now at synthetic fuel production (see Figure 5.6). Even with the vast amounts of money being spent on synthetic fuel projects, it is unreasonable to expect that they will have a major impact on total supply in this century. However, it does not mean that synfuels will not play a critically important role in the supply of oil at the margin. Standard Oil's forecast calls for 4.3 million b/d of synthetic liquid fuels by the year 2000. This volume will come approximately 30 percent from tar sands, 30 percent from coal, 25 percent from shale, and the rest will be alcohol. This amount could vary

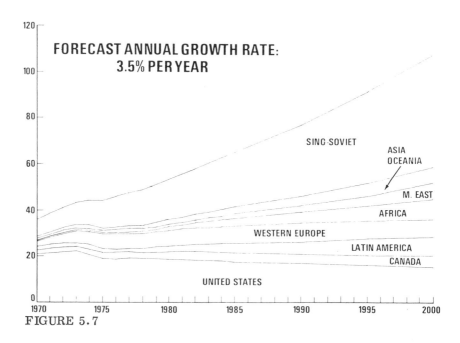

FIGURE 5.7

World Natural Gas Production
(trillion cubic feet per year)

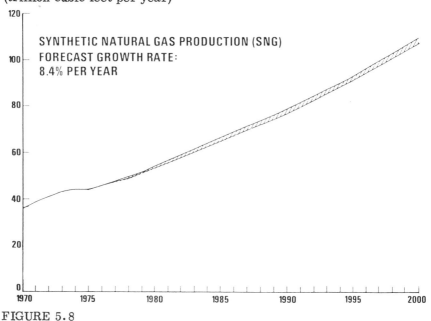

FIGURE 5.8

World Synthetic Natural Gas Production
(trillion cubic feet per year)

144

rather significantly, depending upon the rate at which money is spent on development. As a minimum, at least 2 million b/d will be produced in 2000. On the other hand, it is possible to imagine that world synfuel production in 2000 could be double this forecast, say about 10 million b/d. The most recent EXXON program suggests that such a level of production is possible in the United States alone. However, it is doubtful that this is either reasonable or probable.

A word of caution is in store here. Any incremental synfuels production cannot be considered as a straight addition to the existing oil supplies, as shown here. In fact, in order to generate the activity required to develop synfuels in a timely fashion, the United States will have to experience various disruptions in oil supply during the 1980s with a prospect of substantially lower oil supplies from conventional sources in the 1990s. Therefore, the total supply curve would have to come down in order to generate the kind of activity in synfuels that is required for any significant increase over the base case. This kind of a political consensus in the United States is quite far from being realized at this point. In other words, synfuel production could approach these levels only if there are reoccurring supply shortages in the 1980s that result in a change in basic attitudes toward the need and urgency for this kind of development. As a final note, even the base case forecast represents a substantial commitment to synfuels production beginning in 1980, and this is becoming obvious.

Natural gas is another major source of energy for the world, and it is also one of the brighter spots. Based upon current and future reserve estimates, world gas production will continue to grow by about 3.5 percent per year until the end of the century (see Figure 5.7). This means that natural gas production will double in the next 20 years. Although reserves will easily support this kind of growth, continued development of the distribution system, which in turn depends on stable world conditions, is essential to make this increase in supply a reality. With the exception of the United States, consumption growth will be fairly even distributed around the world, if distribution grids are extended to permit increased use of natural gas.

At the margin, total gas consumption will depend upon the development of synthetic natural gas (SNG), mainly from coal (see Figure 5.8). SNG could amount to around 2 trillion cubic feet a year in 2000, with about half produced in the United States, although the same trends that are seen here could accelerate SNG production in other coal-producing industrial countries, in Europe, for example.

Also at the margin and essential to continued, increased availability of gas in the United States and around the world would be an increased emphasis on liquefied natural gas (LNG) (see Figure 5.9). Only in this way can the world tap the large reserves of natural gas located in the producing areas—Indonesia, the Middle East, Africa—

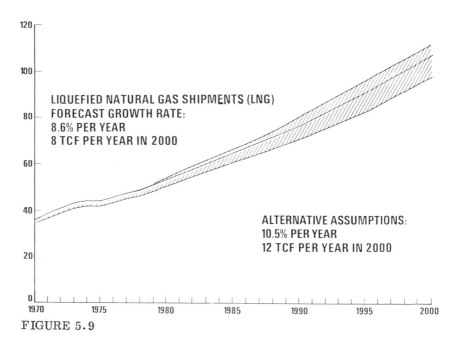

FIGURE 5.9

World Liquefied Natural Gas Production
(trillion cubic feet per year)

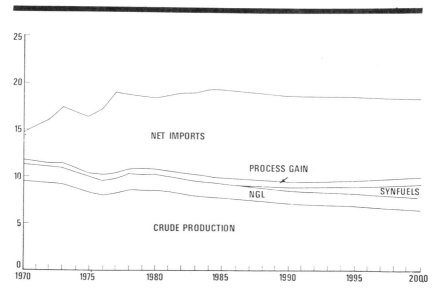

FIGURE 5.10

U.S. Oil Supply
(million b/d)

and bring them to market. Standard Oil's forecast calls for LNG to grow by 8.6 percent per year to a total of 8 trillion cubic feet in 2000. To put this volume in perspective, consider that it is almost one-half of the current U.S. consumption of natural gas. It will be no small achievement to provide the facilities to produce, liquefy, transport, and distribute this amount of gas. It is clear that the gas is potentially available if it can be moved to market. Right now, it seems that a combination of pipelines for shorter hauls across the Mediterranean and perhaps even from the Middle East to Europe, and LNG, for longer hauls, will be used to move gas from producing areas to consumers.

Based upon reserves, LNG movements could be increased from these forecast levels. However, one has to keep in mind just how much LNG projects are going to cost. Current experience indicates that just the LNG portion of the project (liquefaction, transport, and regasification) will probably cost in the range of $10 billion per trillion cubic feet per year of production. On top of this, it is going to be expensive to find and develop the gas fields. The high cost of these incremental gas supplies will definitely be a significant factor in determining demand for gas at the margin. And this, of course, comes back to the fundamental question—is gas a premium fuel or is it not?

The Standard Oil forecast of U.S. oil supply was made earlier in the year when the impact of the 1980 recession was probably not as well known or was not as clear as it is now (see Figure 5.10). It was anticipated that total oil consumption would continue to grow through 1985, after which it would begin to taper off. It may well be true that, in the United States, oil consumption had already peaked in 1978. That is definitely true for gasoline. The United States will never again use more gasoline than it did in 1978. And that may also be true for total oil. This is going to occur because the forces have already been set in motion to cause this. Those forces include higher prices, education, energy awareness, and, most importantly, the expectation of future real price increases, all of which will continue to have a growing effect on demand. The trend toward more efficient use of energy, and especially oil, will be reinforced by market forces and by political uncertainties.

Synthetic liquid fuels are going to grow rapidly during the 1990s. However, before 1990 this will not be seen very much because most of these projects have about a ten-year lead time and are just getting started. Expected are about 1.6 million b/d in 2000. This will, however, be barely able to offset the natural decline rate of conventional oil production. Oil from shale will provide about two-thirds of the synthetic oil production in the United States. Some acceleration of this effort is still possible but seems unlikely to have much of an effect until quite late in the forecast period.

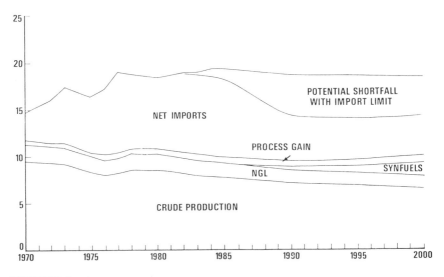

FIGURE 5.11

U.S. Oil Supply and Potential Shortfall
(million b/d)

One final point and one that does not require much comment is
that the basic conclusion of this forecast is that the United States will
continue to be dependent on oil imports at least as far as one can see
into the future (see Figure 5.11). Remember, the demand forecasts
already include some optimistic estimates of conservation potential
and moderate estimates of economic and energy growth rates. Even
so, the United States is showing import requirements 10 percent
higher than the 1985 goal expressed by the administration and double
the longer range goals for 1990.

It should be obvious from Figure 5.11 that the United States has
a long way to go to meet those goals. Incidentally, Ronald Reagan is
now fourth in a line of presidents—starting with Richard Nixon de-
claring Project Independence, followed by Gerald Ford, and then
reinforced by Jimmy Carter's energy programs—to indicate intentions
to make the United States independent of imported oil.

As mentioned earlier, it is important to consider some unfavor-
able facts. One bit of modern-day heresy is that markets do work.
But the effects of moving toward market prices for oil are becoming
evident now that price decontrol is under way. World energy prices
are finally beginning to be reflected in the United States. Now that the
United States has reversed its policy of subsidizing its energy consumption, perhaps it can be a leader in encouraging other countries

who are in the same situation. Indonesia, Canada, Mexico, Saudi Arabia—all of these countries that are major oil producers today—are following the same shortsighted domestic pricing policies that the United States has had for a long time. There is a lesson here that could be passed on to these other countries.

A case can be made for the fact that the financial markets have worked. Certainly the problems now in recycling petrodollars are larger than they were in 1973-74; however, they were regarded then as just as insurmountable as the problems seem today. And yet those petrodollars were recycled quite effectively the last time around.

There are some contingencies that the United States might have to face here. Two types of contingencies must be considered. One is the longer-term scenario, which is depicted in the Standard Oil forecast, that calls for higher energy prices, slower economic growth, slower and slower energy growth. This future is already here; and the United States is already beginning to adjust to it. The transition period was compressed by the Iran revolution, but the United States has actually come through in reasonably good shape.

The other, more difficult problem lies in the area of short-term, abrupt disruptions. The United States is having considerably more difficulty dealing with that kind of contingency. Disruptions can arise from politics, sabotage, natural events, or even major wars. As the world approaches the natural supply limits, whatever they happen to be and whenever it happens to occur, the balance becomes more delicate. When maximum productive capacity is reached—and that was part of the problem in 1973 and also part of the problem at the end of 1979—any disruption becomes more critical. In 1973 it took concerted action on the part of the Arab oil suppliers, and in 1978-79 it took a major shutdown of the fourth largest producer of oil in the world in order to send a ripple through the world oil supply system. In the future, the same effect could be generated by a similar disturbance in any one of a number of considerably smaller, marginal oil suppliers to the world. It can be stated with certainty that there will be a number of Iranian-type crises during the next 20 years, in view of the political instability in many of the oil-producing areas of the world. And this is the reason why more efficient use of energy—conservation, if you will—is essential. Although it is true in the abstract that even a small disruption in energy supply to a very efficient economy may have a larger impact because the "fat" is gone, it certainly does not make any logical sense to continue to be profligate and not worry about conservation at all, just to have some "fat" to trim off in an emergency. It is essential to spend time working on the conservation and the improved efficiency that will come about primarily through the normal operation of market forces.

This discontinuous oil future should result in oil prices going up in cycles rather than continuously. The Standard Oil forecast is that real oil prices will go up about 3 percent per year on the average. However, price increases are more likely to occur as they have in the past; they will go up abruptly, and then taper off or perhaps even go down, as they did between 1973 and 1978, until the next time that the demand curve approaches the supply curve. Then, something will occur to disrupt that stability, triggering another round of price increases.

It is also essential to point out that contributions from all sources of energy are important. For example, I look at residual fuel oil as cracker feed stock that can be made into gasoline. A barrel is a barrel, a Btu is a Btu. All of these forms of energy are fungible to a great extent. We may not have in place all of the equipment required to make these substitutions today, but, in the longer term, we have got to continue to make our use more efficient. We have to substitute where substitution is possible. You cannot run your car with coal; you can run your electric power plant with coal. Put gas into your house and put fuel oil into a cracker to make gasoline to run your car. These sources are interchangeable via the refining and the distribution systems, and we have to increase the flexibility that permits us to use those Btu's, those barrels of oil, those cubic meters of gas and tons of coal, in the most effective way possible.

The bottom line is that we, as a nation, are going to continue to be dependent on foreign oil throughout this century. This means that our energy future is inextricably linked to the rest of the world. Even if we were able to reduce this dependency by increasing domestic production, developing alternatives, by using the oil we do have available more effectively, we will still be unable to sever the link completely. And anyone who tells you differently is not facing up to the reality of the situation.

6

IMPLICATIONS OF OIL MARKET CHANGES FOR FEDERAL AND STATE ENERGY SECURITY

Thomas L. Neff

Rapid changes in oil market conditions raise serious questions about the costs and security of energy supply. In the long run, oil price increases threaten economic health and political stability. In the short run, disruptions in supply and rapid price increases can have even more significant impact. These problems affect all consuming regions and both classes of problems call for new policy responses. The character of the most successful policy responses must depend on the nature and behavior of the world oil market. But this system is changing rapidly and in ways that will make it difficult to deal with energy and security problems, especially those associated with sudden disruptions. As we shall see, a critical issue raised by these changes is the extent to which any given consuming region will be able to deal independently with its energy problems. Indeed, it is likely that some independent efforts, especially if they are pursued by many consumers, will only make things worse. An individual state's welfare is increasingly and unavoidably bound up with national and international market developments and policies. For that reason, an understanding of present market realities and trends is essential to the formulation of appropriate energy security policies for any state.

OIL MARKET CHANGES

Prior to 1973-74 most oil was produced, exported, traded, and its products distributed by the seven international oil companies and other crude-producing companies. About 90 percent of the oil produced in the Organization of Petroleum Exporting Countries (OPEC) was handled in this way. The volume that the companies handled was in fact greatly in excess of the needs of their own systems, making possible about 7 million barrels per day (b/d) of sales to third parties. The oil companies also had agreements with each other to trade oil back and forth to match the needs of their customers; up to several million barrels per day were traded in this way. This system had two important capabilities. First, it was fairly efficient, logistically, and second, there were intermediaries that stood between consumers and producers. The producers had relatively little downstream control over what happened to their oil. This was very useful in the 1973-74 crisis; although the disruption of supply was targeted against particular consumers, the system was able to reequilibrate and to reallocate oil supplies internationally. What would have been very large shortfalls for one or two consumers became small shortfalls for everyone.

The changes initiated by the events in Iran have altered this situation considerably: we have lost much of the flexibility the international system had to adjust to crises. Following the Iranian revolution, exports from that nation dropped suddenly by about 5.5 million b/d; the majors lost a little more than 3 million b/d. When Iranian production increased again, only about 1 million b/d of the increase went to the majors. What altered the working of the international oil market was not simply the absolute quantitative loss in oil supplies, but rather that the suddenness of the change caused consumer panic and a competitive scramble for remaining oil supplies. Consumers felt a strong need to increase inventories, so demand went up rapidly. This disruptive atmosphere created the opportunity for other producers to start imposing conditions on the supply of oil—something they had long wanted to do.

In the post-Iran market, two trends are evident: producers are taking greater physical control over their crude oil and its uses, and they are extracting higher economic and political prices for that oil. It is notable that there are significant differences among OPEC producers in the extent to which such trends are manifest, and also that some non-OPEC producers (for example, Mexico, Peru, and even Norway and the United Kingdom) are taking actions very similar to those of the OPEC producers.

With these market changes, the role of the international majors has been altered significantly. Producers are now selling much more oil directly to other oil companies, to consumer governments or their agents, or on the spot market. Direct sales have increased to nearly 50 percent of the producers' total volume, up from about 10 percent in 1973. As a result, the majors have essentially been cut back to their own systems. Third-party sales are declining, exchange agreements are breaking down, and the system is becoming increasingly inefficient.

Today, about 13 million b/d of internationally traded oil (both OPEC and non-OPEC) are being sold directly by producers. Roughly half of this oil is being sold in transactions between producer and consumer governments, usually at the official government selling price. Over much of the past two years, this has been well below the price for other direct deals that often include a premium of as much as $11 per barrel. In many government-to-government deals, however, the buyer agrees to provide technological or other assistance. In other cases, the "premium" is less explicit: typically, political concessions that rarely extend beyond "goodwill." But the establishment of any political context for oil transactions could lead toward serious constraints on consumer actions, and could reduce the flexibility of the oil market considerably.

A similar volume of crude is being sold directly to independent oil companies, traders, trading companies, refiners and distributors, and the national oil companies of consumer countries. The independent companies, with their new and expanded roles, may be extremely important to the functioning of the international oil market under both normal and abnormal conditions.

Under both government-to-government and other direct transactions, the flows of crude are often subject to important new restrictions. System restrictions require that the crude be used in the refineries of the purchaser; resale restrictions mandate that the oil and its products be used in the home market of the purchaser; destination restrictions prohibit, for example, shipping through the Suez Canal. Other types of restrictions now being imposed by producers include anticompetition clauses; mandatory purchases of petroleum products or lower-quality crude as part of a "package deal" with more desirable crudes; exploration requirements (or fees to finance exploration); and requirements that crude be transported in tankers owned by the producing country.

The result of all these new restrictions is a great reduction in the flexibility of the international system to respond to crises. In normal times, there is also a loss, and that is a loss of economic efficiency. A very constrained market means more costly transportation, more refinery and crude mismatches, and other inefficiencies. But in a crisis the problem is potentially much more serious. The danger is that such restrictions will inhibit the ability of the international market to respond to disruptions by redirecting oil flows.

Producers have also used their new crude supply leverage to create opportunities for participation in transportation, refining, and other downstream operations. Some are building refineries or petrochemical plants at home or abroad, often in joint ventures with foreign oil companies for whom such investments are a condition for access to crude supply. Others are asking for a share of downstream profits.

And at least one producer, Kuwait, is asking for the right to refine a quantity of its own crude in the purchaser's facility for its own account, rather than to the benefit of its purchaser. The motives for these steps seem to include downstream experience, products for home markets, control over or information about oil companies, or simply capturing downstream profits. However, such participation has risks; producers may find it safer to continue to keep crude prices up and to rely on middlemen to absorb losses when, for example, there is a product glut. Producer movement into refining and product markets may prove to be a self-limiting phenomenon, with considerably less effect on international oil trade than will result from their enhanced control over crude itself.

Changes in price structure have paralleled the increasing complexity of oil market structure. In the past, oil was sold almost entirely at official government selling prices. Price variations were due largely to quality and transportation differentials, and price unity among the members of the OPEC cartel was readily maintained. With the events in Iran, producers were suddenly able to act independently to alter pricing practices. Oil is now sold in three primary categories: long-term contracts at the official price, term contracts at the official price plus a premium, and single cargoes or short-term sales at spot prices. Official prices are generally charged for the "equity" oil sold to major oil companies and for many government-to-government transactions.

Other term contract transactions, totaling perhaps 30 percent of OPEC sales (in early 1980), are generally at the official price plus a premium. The premium may take many different forms. For example, Kuwait simply added a $5.50 per barrel surcharge on quantities in excess of a basic volume for each purchaser; Iran required that crude purchasers also buy fuel oil; Mexico required that buyers take heavy as well as light crude; and Algeria collected a $3 per barrel fee to finance exploration. Variations in such premiums have been greater than official price differentials; that the market could sustain such differences is a measure of the disturbed state of the international oil system.

Spot sales are an even more sensitive measure of market instability. In the scramble to rearrange oil supply following the disruptions caused by events in Iran, producers found that they were able to sell large volumes of oil on a spot cargo basis at high prices. Previously, the spot crude market had involved relatively small quantities of oil sold by middlemen in Rotterdam, Singapore, or elsewhere; spot sales were basically a mechanism for clearing the market at the margin. Producers thus created an essentially new spot market in 1979, a market of greater size and with a different role in the oil system. Fluctuating with the precise state of the market and with consumer perceptions, volumes and prices for such sales have varied greatly over the past year. In mid-1979, spot prices were as much as double official prices on a volume of several million barrels a day. More recently they have declined, approaching premium prices for term oil on a volume of less than 1 million b/d.

This recently evolved three-tiered price structure is important not simply because of the numbers associated with the prices, but because this structure allows great volatility and numerous mechanisms for raising prices. While approximately two-thirds of all oil is still sold at official prices, the ability to sell large cargoes on the spot market or large amounts of oil at a premium above official prices puts a very strong upward pressure on those official prices, particularly when the market is tight. Official prices apply primarily to "equity" oil sold to major companies and in government-to-government deals. A producer selling to an oil company has an incentive to reduce the company's share of production in order to exact premiums in direct sales or political concessions in government-to-government deals.

The effects of all of these new structural rigidities and inflexibilities have been aggravated because the average barrel produced is becoming increasingly heavy and high in sulfur content. At the same time, the average barrel demanded is shifting toward the light, low-sulfur end. As a result of this—and the limited capabilities of current refining capacity—the efficiency of the world system in matching crude supplies to final product demand is reduced further and its capabilities for substituting and reallocating oil in a crisis are further compromised.

DIRECTIONS OF CHANGE

The many changes under way in the world oil market will inevitably alter how we think about the problems of oil and security, and how we deal with them. But the difficulty of our present circumstances is that we cannot yet see the full extent and implications of these changes. One could imagine two extreme scenarios for the possible evolution of the market. In the first, producers would establish complete control through destination restrictions, government-to-government agreements, downstream involvements, and tight leashes on the majors and other market participants. Such a market would have few degrees of freedom left. It would be highly politicized and economically inefficient under normal circumstances: In a crisis, it would be inflexible, prices would be volatile, and readjustment costs would be very great.

In the second scenario, producers would also succeed in removing much of the flexibility of the old system, but new actors and market mechanisms would restore flexibility and efficiency to the system. These might include independent oil companies, refiners, trading companies, and expanded and elaborated oil product markets. In such a system, these intermediaries would provide logistic and economic efficiencies. In a crisis, they would help reallocate oil much as the majors appear to have done in the 1973-74 crisis.

Of course, the world oil supply system is not presently in either of these extreme states, nor is it likely to come to resemble either one completely in the near future. Flexible and inflexible elements will coexist, their nature and relative importance altering with time.

To understand the system and its potential evolution, it is useful to consider the analogy of a pond that is close to the freezing point. When water freezes, many degrees of freedom are lost; fluidity is replaced by rigid relationships between the parts. A liquid has many more degrees of freedom: a pressure applied to any part is distributed over the whole fluid system and disturbances die rapidly away in ripples.

For the world oil market, the frozen state is analogous to the first of the extreme evolutionary possibilities above. In this situation, it is likely that stresses could be directionally effective against particular consumers (as in a targeted embargo) or cause major disruption in the entire system. But such a rigid system might also have certain advantages for consuming nations. For example, producers may have less freedom in individual actions or in their ability to act together in a cartel because of government-to-government agreements or investments in downstream activities or consumer nations' economies. A fluid market has more evident advantages; its disadvantage is probably just the lack of constraints on producer behavior.

The current world oil market is clearly in a mixed state. There is evidence of both freezing and thawing, both ice and open water. Some degrees of freedom, formerly provided primarily by the majors, are now frozen out by supply reductions and by the growing number of direct deals and special restrictions; but other sources of flexibility may be emerging. The difficulties of our situation are, first, that we cannot yet tell what the basic direction of the system is—whether it is spring or fall—and second, that many of our previous understandings and analytic approaches to the world oil market are not adequate for a system whose basic state is changing. But it is important to identify the forces at work and trace their implications, since the policies and actions of the participants in the world oil market can alter the evolution of that market.

The expansion of government-to-government agreements and the wide range of other arrangements in which the particular interests of individual producers and consumers are expressed tends not only to reduce flexibility but also to reduce cohesion among producers and among consumers. The tightness of the market, the high level of supply-security concerns among consumers, and the rise of producer revenues well above expenditures enables producers to act more independently in setting prices, production levels, and other terms. Decisions and events in many producer countries can result in major short-term perturbations in the system and affect its long-run evolution. This new volatility on the producer side is likely to lead to an unpredictable variance in market behavior, including prices.

Fragmentation and volatility also affect consumers. Government-to-government transactions, for example, may induce a consumer to sacrifice nonoil domestic and foreign policy goals in the belief that this will ensure supply security or more acceptable prices. Each consumer will generally perceive a different set of such goals, and weigh trade-offs differently. This fragmentation of perspectives compromises consumers' efforts to cooperate in dealing with oil problems; it also increases the potential for significant conflict among them.

Consumer reactions to perceived insecurity have also contributed to the problem. Events in Iran catalyzed market changes not so much because of a decrease in total supply but because of the panic induced by uncertainty and by the need to rearrange supplies. Consumer and company responses enhanced volatility and tightening of the market, which in turn set the stage for other producer actions. Measures to restrain or reduce the effects of such consumer responses would help inhibit such market changes in the future. However, given the changes in the market that have already occurred, consumers may be unable to achieve the coordination and sharing required to implement such measures.

Not all of the trends toward structural rigidity and fragmentation necessarily work to the disadvantage of consumers. The government-to-government ties that restrict consumer action may also restrict producers: the chains bind at both ends. And since producers have disparate interests, the growing web of bilateral relationships may make it more difficult for producers to act in concert. In addition, the existence of a particular bilateral relationship may reduce the likelihood of disruption of the corresponding supply channel, or at least this is the hope of the consumers involved. But if there are many such agreements and one disruption occurs, that consumer may find its ability to arrange new supplies limited by the government-to-government deals of others. Thus, a measure that appears to enhance supply security for individual consumers may also serve to rigidify the system further, making disruptions much harder to deal with when they do occur.

While there might be some consolation to be found in the market's trends toward rigidity, it would be more encouraging to find indications of changes that help restore the flexibilities and fluidities that have been lost. There are indications of such changes, but their significance is not yet clear. As discussed earlier, some of the oil taken from the majors and other participating companies is being sold, not in government-to-government deals, but to independent oil companies, traders, and trading companies, some of which have shown increasing capabilities in the complex international transactions that are necessary to provide increased flexibility. But the volume that is being traded internationally in this way is still far from comparable to that previously handled by the majors as logistic exchanges or third-party sales, and producers will probably resist the replacement of the majors with an even less controllable group. Indeed, several producers are now putting restrictions on the sizes and types of companies to whom crude is sold and setting lower limits on sales volumes.

THE NATURE OF CRISIS

The implications of these changes in the world oil market are troublesome under "normal" market conditions (if today's market is regarded as the new form), especially as regards price and efficiency. Of even greater concern are crises of supply. We have experienced two such crises in the past decade, and the prospects for further crises seem only to be increasing. It is useful to distinguish two classes of events: intentional interruptions (such as an embargo) and accidental or undirected disruptions.

Intentional interruption is likely to be motivated by political objectives and targeted against particular consumers. Under earlier market conditions (as in 1974), it was difficult for producers to carry out an embargo successfully, since the majors and other intermediaries could reallocate supplies. In today's altered market it is not clear that there are enough flexibilities in the international system to cope with intentional interruptions. But market changes may also be working to decrease the likelihood of such intentional interruptions. Producers can now use both price and access as tools to advance political and other goals, and each producer can act independently to achieve its goals. If an embargo were targeted against a consumer that had government-to-government agreements with other producers, these producers would have to abrogate these agreements in order to make the embargo successful. Not only do individual producers now have other sources of leverage, but their ability to act in concert has been reduced by the very market changes that otherwise would increase the effectiveness of an embargo.

Future crises are therefore more likely to result from accidental disruptions than from embargoes. War, revolution, natural disaster, or other causes may reduce total world supply and initiate new scrambles to reallocate what is available. Such a cut may be relatively small—say, the loss of a few million barrels a day—or large, as with the loss of Saudi Arabia or several smaller producers simultaneously. For large disruptions there is little hope of avoiding very serious political and economic costs. Even for smaller disruptions, costs may prove larger than in the past. Increased structural rigidities inhibit the nearly automatic reallocation of oil that took place in the past, and this inhibition may be greater than any new fluidities introduced by other market changes.

During a disruption that had no overt political motive, some producers might be willing to increase total supply enough to overwhelm logistic tangles, meet new demand for inventories, and put a lid on prices. Some might also relax destination and other restrictions temporarily. But producers are likely to exact economic and political concessions in return for such favors. Moreover it is not obvious that adequate logistic infrastructure for rapid large-scale reallocation, or for the handling of additional quantities from other producers, would be maintained simply as a contingency.

On the consumer side, governments could reallocate oil, restrict demand, and distribute economic costs. Particular attention has been given to sharing agreements as a way for consumers to deal with disruptions. The difficulty of implementing such agreements has been increased by recent changes: new efforts and commitments will be necessary if sharing agreements are to be of value in a crisis. In many cases, reallocation of one consumer's supply to another consumer might now violate (or appear to violate) government-to-government or destination restrictions, and consumers might well fear that such violations would compromise future access to crude. Finally, the efficacy of such sharing arrangements depends on consumers' abilities to reallocate oil. In earlier years, the international majors would have done the reallocating; now, governments may have very limited capabilities for reallocating oil quickly and without large economic costs, even if the political will can be marshaled.

RESPONDING TO CHANGE

The evolution of the world oil market can be affected, for good or ill, by a number of consumer policies and actions. At a minimum, consumers should avoid measures that reduce system flexibility (such as emergency allocation systems that misdistribute oil or products). The test of flexibility should be applied to all energy-security policies; a primary consideration for any proposed measure should be whether it increases our ability—locally, nationally, and internationally—to respond to crises events. We can distinguish three classes of constructive measures: those that increase flexibility directly; those that augment supply, especially in time of crisis; and those that reduce demand. The last two operate primarily by enhancing opportunities to exploit existing flexibilities.

Some new flexibilities may emerge automatically, as in the expanding role of new actors in the market. Attempts by producers to dictate the behavior of one small buyer may merely result in the appearance of another. By this argument, consumer policies that encourage the active participation of such entities in the international market will help increase flexibility. Some current policies encourage such participation; others (such as the crude allocation system that tempts smaller companies to depend on larger) may discourage it.

The recent trend toward holding larger crude and product inventories may be another beneficial development. Such inventories provide a cushion of time; they can restrain the hoarding impulse that turns a disruption into a crisis; and they can also provide the flexibility to redirect crude streams temporarily, or to reallocate products. To do this, however, inventories must be held in ways that mesh well with domestic and international distribution systems. Since these systems are largely private, reallocation might take place more promptly and efficiently if inventories were privately held. But since the societal value of such inventories is generally greater than their value to private agents, private holders will probably maintain smaller inventories than are socially optimal, unless there is some public intervention in the form of regulation or subsidy, as there is in Japan and some countries of Western Europe.

Product markets have also expanded and probably should be encouraged to do so further. Products (with the exception of unleaded gasoline) are far more easily exchanged than crude oil, with its many quality differences and with the variety of producer restrictions attached to its use. Although the costs of storing and transporting products are greater than those associated with crude, the ability to respond quickly and flexibly in a crisis to meet product demand could help consumer nations minimize economic disruption and panic about supply.

Besides these trends toward flexibility, which can and should be encouraged, there are a number of positive steps that might be taken by consumers. Perhaps the most important technical measure to increase flexibility is the upgrading of refinery capacity. Many refineries can handle only certain types of crude oil, usually tending toward light low-sulfur crudes. There are too few ways to match refinery capabilities with the crude slates that might be available in a crisis. This constraint multiplies the effect of producer destination and other restrictions. For example, the loss of a supplier of light low-sulfur crude probably could be compensated for in the United

States only if other consumers agreed to accept less desirable crudes (and the associated environmental costs—e.g., increased SO_2 emissions) in order to free high-quality crude for use in U.S. refineries. Although private companies may find it profitable to upgrade refineries (particularly if lower-grade crudes are disproportionately low in price), such incentives do not completely reflect the societal benefits due to increased flexibility and ability to respond in a crisis.

Measures to increase other logistic degrees-of-freedom would also be worth far more than their immediate costs, but such measures are sometimes difficult to achieve because they involve other public policy issues. For example, Alaskan oil is politically restricted and does not play a role in international oil trade. To a more limited extent, this is also true of Mexican oil (which comes primarily to the United States). Such restrictions may be politically justifiable and perhaps economically efficient in normal times, but a failure to provide for the flexible use of non-OPEC oil in international markets probably increases the impact of OPEC constraints during crises. For example, a crisis affecting light, sweet crude supplies might be especially damaging to the United States unless heavy Mexican or North Slope crude could be sent to refining centers in the Far East or Europe in order to displace premium crudes for use in U.S. refineries. Restrictions on transfer of non-OPEC oil may seem to enhance the supply security of the countries who consume it, but in fact the flexible use of non-OPEC oil is very important to the ability of the international system to cushion shocks and minimize their effects.

Market flexibility would also be increased by further diversification of supply sources. The importance of new sources can be much greater than is indicated by the volumes involved if they contribute to increased flexibility in the system. Even small producers, or small increases in the capacities of established producers, can alter system performance, particularly if such producers are willing to act independently.

In addition to efforts to increase flexibility directly, there are also measures that would augment supply or reduce demand. In a crisis, such measures can reduce market tightness and enhance the effectiveness of system flexibilities. In normal times, such measures might even influence the evolution of the world market in constructive ways.

The most desirable augmentation of supply would come from the creation of additional production capacity in stable and politically neutral areas. However, it is difficult to identify sources that are large enough to affect the prevailing supply/demand balance appreciably, especially in a world in which producers with revenue surpluses are able—indeed eager—to cut output. And for economic and political reasons, new producers would probably rather sell into a tight market than help contribute to its softening.

However, even if it is not possible to find additional sources large enough to affect market developments, some producers might be willing to install capacity in excess of current needs, to be used in a crisis. At present, much of this surge capacity is in Saudi Arabia. While seeking to make this source of emergency capacity more secure, consumers might also seek to convince Mexico, Norway, or other producers to create surge capacity. However, it is difficult to implement the complex multilateral agreements necessary to achieve a reliable surge capacity outside consumer borders. Some consumer nations with domestic oil fields, such as the United States, could install additional production capacity for use in a crisis, though this temporary use might—for technical reasons—compromise long-term production from the fields affected. Because reliable sources of surge capacity do not exist, a number of consumer nations have sought to augment crisis supplies by increasing domestically held stocks. However, the acquisition of stocks can create economic and political strains. The new demand tightens the market, leading to upward pressure on prices and perhaps even helping to set the stage for the disruptions that one is trying to avoid.

In order to acquire strategic stocks and use them effectively several market-related issues must be addressed. Stocks must be accumulated and distributed through the market channels that ordinarily distribute crude oil and products. A separate emergency system would be costly and unreliable. In addition, a workable plan must be devised for making and implementing decisions about stocks—how and when they are to be purchased, how they are to be priced and allocated, and when they are to be used. In the ordinary market, overall stock policy is the sum of many small decisions made by private actors. As indicated above, this may result in normal stocks that are smaller than society might find desirable on security grounds and too much demand that stocks be increased rapidly in moments of crisis. Thus, while it is possible for the market to make and implement decisions about stocks, private decisions may not be socially optimal. There seem then to be two alternatives for national policy: to find mechanisms to influence private stock decisions in socially beneficial directions, or to accumulate and hold stocks publicly. Several nations have taken the first course, working through regulation and subsidy. At present, the United States has chosen the second. But public stocks require that governments make and implement highly visible, politicized, and costly decisions. It is not obvious that this is possible, as the United States has discovered with its strategic reserve. And although it would be desirable to develop both public and private stocks, the effort required to pursue public stocks in the United States appears to have left little political capital to devote to private initiatives.

It is also very important to realize that stocks cannot be considered solely in a domestic context. The oil market is international, and the United States and other countries act within a broad web of economic and political interdependencies. When reserves are built or used, they flow out of or into this international system. In a crisis affecting many nations or regions, those holding stocks would be under pressure—or explicit obligation, as in the International Energy Agency (IEA) sharing agreement—to use them to meet as much of the total shortfall as possible. Thus from a narrow perspective, the value of a strategic reserve can be diluted by the needs of neighbors and allies. Ironically, consumer fragmentation and losses in market flexibility may tend to decrease the international role of stocks and make them more a matter of domestic security alone. But any rational attempt by one consumer or region to enhance security through stock-building must take account of the role those stocks would play, explicitly or implicitly, in the larger system.

Demand reduction is the other strategy proposed as a response to present market trands and security risks. Obviously, any reduction in the dependence of a nation on particular supply sources lessens its direct vulnerability. But given the interdependencies of markets and alliances, there are important indirect vulnerabilities. For example, a nation or region may not import any oil from a particular producer, but if exports from that producer stop, efforts will be made (by the market and under sharing agreements) to reallocate that nation or region's oil supplies. Consumers might be wise to pursue a diversification of customers for risky producers, in order to minimize the potential loss to any particular consumer country and the high economic and political costs of reallocation should interruption occur.

Long-term demand reduction is often proposed as a way to loosen the supply/demand balance, reduce the upward pressure on prices, and make it easier to cope with crises. If long-term demand reduction would actually accomplish these objectives, it would be extremely important. Unfortunately, it is likely that achievable reductions in demand would simply be matched by reductions in supply, with the market remaining tight. Many producers, especially Saudi Arabia, are producing at levels that generate more revenue than they need, in part because reductions in output could have a catastrophic effect on Western economies (and thus on producer's investments, military security relationships, and so forth). Some of these producers would have an easier time reconciling domestic political forces—which now include strong conservationist voices—if output were lower. These producers would welcome a reduction in demand—indeed several have called repeatedly for such reductions.

There are, however, some better arguments for demand reduction. Reduction in long-run demand may be one way to create surge capacity and increase other market flexibilities. Convincing producers to maintain excess capacity created automatically through demand reduction would be easier than convincing them to create new capacity for use in a crisis. Indeed, some producers might explicitly agree to maintain surge capacity if consumers agreed to curb demand. In addition, output reduction would come primarily from OPEC producers; demand reduction would thus decrease the relative importance of these producers in the world market. The result could be an increased average flexibility in the international system. Finally, reduced output may alleviate internal political stresses in producer countries, lowering the likelihood of interruption of consumer supplies.

Short-term demand reduction strategies, usable in a crisis, might, in fact, be of greater value than long-run demand reduction. Short-term demand reduction has two aspects: in a crisis, one must be able to restrain both consumption demand and the demand for additional inventories. Efforts to increase stocks prior to a crisis can help reduce panic buying during a crisis and thus are an important complement to measures to effect short-term reductions in consumption. These measures range from an emergency tax (which, properly implemented, might also restrain crisis-induced stock building and encourage the use of inventories when appropriate) to long-term shifts in social or economic structure that would allow a more elastic demand response in times of crisis.

As with supply augmentation, the goal of short-term demand reduction is to loosen the supply-demand balance in a crisis. This loosening may be critical to the ability of the system to find and exploit the flexibilities necessary to reallocate oil supplies.

The goal that should guide policy is not import reduction, energy independence, or even domestic energy security: it is flexibility. We are moving into an era in which disruptions seem increasingly likely and increasingly unpredictable as to origin and magnitude. At the same time, the international oil system is becoming more fragile and less flexible. In such a world, true independence or security of supply, through bilateral deals or other measures, is powerfully attractive but an ultimately illusory goal for most nations. Instead, individual and collective efforts should be directed toward policies that increase flexibility and improve the ability to manage crises, whatever their origin. Improvements in oil market flexibility—in its ability to reallocate and reequilibrate during and after a crisis—will not only increase the chance that the market alone will be able to deal with at least some crisis, but will also amplify the effects of emergency measures directly initiated by consuming nations. The ability to smooth the path through times of crisis will help prevent the severe damage we are now all too aware can accompany even small disruptions: the extreme ratcheting-up of prices, the negative impacts on general economic activity and political stability, and the high costs to foreign policy.

Measures to increase flexibility and otherwise to respond to crises will undoubtedly be expensive, economically and politically. But without new initiatives domestically and internationally, there is serious danger that the world oil system will become increasingly fragile and lose what resiliency it still possesses. If this happens, the costs of the next disruption will be even greater than those of the past.

IV

Critical U. S. Energy Sources: Mexico, Venezuela, and Indonesia

INTRODUCTION

The contributors in Part III presented a partial picture of what Tom Burns called the facts about the real energy situation in the world and in the United States—that the United States will continue to be dependent on oil imports as far as we can see in the future. This uncomfortable position is contrary to the rhetoric on energy independence expressed by former President Richard Nixon and all of his successors, including President Ronald Reagan. The public must be even more confused when the media, public officials, and energy "experts" proclaim that perhaps Mexico will be the salvation for our energy needs, or maybe Indonesia, through liquefied natural gas, or Venezuela through development of its extensive reserves of heavy oil.

Part IV outlines clearly the limitations on these possible sources of petroleum and energy resources. Each of the authors points out that internal domestic energy needs, internal political factors, international political issues and responsibilities, and other variables intervene in the overly simplistic energy plans presented to people in the United States by many of its political leaders. The contributors all agree that decisions involving energy sources will be made in Mexico, Venezuela, and Indonesia by officials of those nations and not by public or private decision makers in the United States. As such, depending at least somewhat on the skill and success of policymakers, those decisions may or may not coincide with what the United States desires or feels is in its best interest. Just as we found in Part III that the Middle Eastern nations could or would not solve energy problems in the United States, the chapters in this part reach a similar conclusion in regard to three additional important energy-producing countries.

7

MEXICO'S OIL AND GAS POLICY: PRESENT AND FUTURE

Gary J. Pagliano

To obtain the proper perspective on Mexico's oil and gas policy, it is important to understand: first, how certain nonenergy domestic factors influence its oil and gas policy, and second, how Mexico fits into the world oil and gas picture. It is these domestic and international factors that will influence future Mexican oil and gas production, and it is these factors that will have to be addressed if a sound relationship is ever to be formed.

HISTORY

First, there is the historical factor. Events such as nationalizing Mexico's foreign-owned industry in 1938, forming the state oil company, Petroleos Mexicanos (PEMEX), with its social mandate for providing energy to the citizenry, and surviving the U.S.-European boycott of Mexican oil before and after World War II have brought to Mexico's oil industry social and political significance of the highest order. PEMEX has become a symbol of the country's political and economic independence. Mexicans are intensely nationalistic about their oil, and are wary of foreign countries that are perceived as potential exploiters of their resources.

ECONOMIC PLANNING

Second, there is the economic planning factor. Prior to the early 1970s, Mexico's economy had had 20 years of impressive growth in gross domestic product (GDP), averaging 6 to 7 percent per year in real terms, with relative price stability. When President Lopez Portillo took office in 1976, Mexico's economy had experienced three bad years of inflation, an increasing balance-of-payments deficit, and a plummeting international credit rating, all of which contributed to the 50 percent devaluation of the peso. As part of his plan to restore domestic and international confidence, Portillo and PEMEX announced a large investment plan to develop Mexico's oil and gas resources. The program established an oil production goal of 2.2 million barrels of oil per day (b/d) by 1982, which was later changed to 1980.

Led by the success of its oil program, Mexico's economy rebounded to its previous growth rate of 6 to 7 percent. The growth of PEMEX and industry has been impressive. Industrial growth, led by PEMEX and the automotive, steel, and construction sectors, has increased by about 10 percent annually. Since 1976 oil production has almost tripled, and oil export revenues have increased many times over—approaching $11 billion in 1980.

This success, however, is bittersweet because three major problems remain with Mexico's economy. First, Mexico cannot feed its population. Agricultural growth continues to lag behind Mexico's population growth. People continue to leave the farms and search for jobs in the cities, but most instead join the unemployment ranks in those cities. Agricultural imports contribute to Mexico's second problem: an ever-increasing dependence on imports. Imports can be a measure of the success of a country's economy in meeting the demands of its people. Imports put pressure on the country to keep its balance of payments on an even keel by maintaining a high level of exports; Mexico's main export, of course, is oil. Imports can also take jobs from the importing country; in Mexico's case, that is the most serious consequence of all.

Population growth and unemployment constitute the third and worst problem. Although official employment data are not available, unemployment and underemployment are close to 50 percent. This percentage could increase because Mexico is sitting on a population time bomb. Mexico's population, already 65 million, is expected to double by the year 2000. Currently, 750,000 new workers enter the job market annually, but only half of them can find jobs. Many analysts argue that population growth is not only Mexico's worst problem but also the least subject to an early cure through increased oil revenues. This is especially true if oil revenues are reinvested in capital-intensive industries instead of being used to develop more labor-intensive sectors.

TRADE

There also is the trade factor—particularly trade with the United States. Mexico's proximity to the United States permits fast delivery, low transportation costs, and easily accessible servicing and technical assistance for both countries. About two-thirds of Mexico's total trade either comes from or goes to the United States. Mexico wants to diversify its trade markets, mainly in order to detach its economic future as much as possible from that of the United States. Mexico has chosen the oil market as the main avenue for diversification. In 1977 at least 85 percent of all Mexican oil exports went to the United States, with the remainder going to Israel and Spain. By the end of 1980, only about 60 percent of the oil exports will go to the United States; the remainder will be distributed to more than 10 other countries (see Table 7.1).

TABLE 7.1

Approximate Distribution of Mexico's Allowed Oil Exports at the End of 1980
(b/d)

Destination	Volume
United States	655,000
France: CFP	100,000
Spain: Hispanoil	100,000
Japan	50,000[a]
Israel	45,000
Brazil	20,000
Jamaica[b]	16,000
Costa Rica	7,000
Nicaragua	7,500
Barbados	3,000
Dominican Republic	14,000
El Salvador	8,000
Guatemala	8,500
Honduras	6,000
Panama	12,000
Canada	10,000

[a]Volume to rise to 75,000 b/d in third quarter and 100,000 b/d by end of 1980.

[b]Starts July/August.

Source: Petroleum Intelligence Weekly: The Journal of Commerce, New York, N.Y., May 5, 1980, Vol. 19, No. 18, p. 3.

It should be noted that absent from Table 7.1 is the additional 200,000 b/d of oil that Japan seeks in return for providing economic and technological aid in Mexico's industrialization. Also on an "if and when" basis are the letters of intent to supply 70,000 b/d to Sweden and 40,000 b/d to Canada, signed during political talks with Mexico.

Mexico's export commitments somewhat exceed current exports. If Mexico is to meet all of its oil export commitments, it will have to increase production and its exports. In September 1980, President Portillo announced that Mexico's production goal, or "platform," would be raised to 2.7 million b/d by early 1981, or about 20 percent over the 1980 figure.

FOREIGN POLICY

Mexico's oil trading partners are diverse in their economic and political character. They include industrialized countries such as the United States and France as well as less-developed countries (LDCs) such as Costa Rica and Nicaragua. Mexico's international energy policy has been to track OPEC world oil prices, bargain with the industrialized countries—exchanging oil for technology and nonenergy trade concessions, and to make special arrangements with certain LDCs.

This policy is part of President Portillo's world energy plan outlined before the United Nations in September 1979. The plan calls for the industrialized and major energy-producing countries of the world to ensure that all countries have an opportunity to meet their energy needs. The plan's major elements are to: first, guarantee the full and permanent sovereignty of each nation over its own natural resources; second, increase the systematic exploitation of potential conventional and unconventional resources by proving financial and technical assistance; third, establish a short-term policy to address the problems of oil-importing developing countries; and fourth, set up a development fund supported by industrialized and producer-exporting countries to meet both the long-term and short-term objectives of developing countries.

After waiting one year, Mexico and Venezuela embarked on their own version of an international energy program in the Western Hemisphere. Built on a program started by Venezuela in 1975, the two oil exporters agreed to supply 160,000 b/d to the LDCs listed in Table 7.1: Barbados, Costa Rica, El Salvador, Guatemala, Honduras, Jamaica, Nicaragua, Panama, and the Dominican Republic. The oil will be priced at each supplier country's official rates, with 70 percent payable in cash at regular commercial terms, and 30 percent financed by five-year government credits at 4 percent interest. If the credits are used for economic development projects or alternative energy sources, the credits will be available for up to 20 years at 2 percent interest.

The LDC program is just one example of Mexican oil being placed in world trade because of some economic, trade, or foreign policy priority of the federal government. This, of course, is true of OPEC states as well as other non-OPEC states like Norway and the United Kingdom. It will be the economic and geopolitical context just described, coupled with oil and gas technical factors, that determines future Mexican oil and gas policy.

OIL AND GAS RESERVES

In September 1980, Mexico's proven oil and gas reserves were raised from 50 to 60 billion barrels of oil equivalent. Depending on what kind of oil and gas mix ratio is used, this proven reserve figure breaks down to about 40 billion barrels of oil and 108 trillion cubic feet of natural gas per day (cf/d). The oil figure makes Mexican reserves the fifth largest in the world, behind Saudi Arabia (163 billion barrels), the USSR (67 billion barrels), Kuwait (65.4 billion barrels), and Iran (58 billion barrels). The United States has proven oil reserves of 26.5 billion barrels.

The ultimate size of the oil and gas reserve base has been subject to much speculation, partly because PEMEX never releases "hard core" geological information enabling energy experts to make more precise estimates. Instead, it simply periodically announces a new set of reserve numbers. This author agrees with the ultimate recovery estimate for oil and gas prepared by the U.S. Geological Survey (USGS): 70 billion barrels of oil equivalent.[1] The USGS analysis is based on an evaluation of all available published and unpublished data, as of January 1, 1980. A Rand study assigned a 90 percent confidence level to a similar estimate.[2]

MEXICO'S PRODUCTION CAPABILITY

In January 1979, a Congressional Research Service (CRS) report made a forecast of Mexico's crude oil and gas production to the year 1988.[3] The report assumed Mexico would reach its 2.2 million b/d platform by 1980, and estimated that if there were gas exports, presumably to the United States, crude oil production could reach 3.8 million b/d by 1988.[4] With Mexico currently producing 2.3 million b/d, and reaching 2.7 million b/d by the end of 1981, it is conceivable that the platform could reach 3 million b/d by the end of 1982. Any projection beyond 1982 was considered speculative since only the next president can change that platform.

Under President Portillo, Mexico has invested heavily to expand its domestic refining capability. PEMEX has spent $950 million in new plants over the last two years and will spend another $1 billion by 1982. The new plants have already increased refining capacity 12 percent annually to over 1 million b/d and further investments should expand capacity to 1.7 million b/d by 1982.

Natural gas production is currently 3.5 billion cf/d. The CRS predicted that production of associated gas and nonassociated gas would reach 8.4 billion cf/d in 1988. Accommodating higher gas volumes depends on Mexico's domestic demand and its potential for exporting gas.

Mexico's gas deposits are relatively concentrated and subject to collection in pipeline systems. There are virtually no remote gas deposits.

MEXICO'S DEMAND FOR OIL AND GAS

As part of its original 1976 oil and gas plan, Mexico decided to continue to rely on oil as the mainstay of future domestic energy development and to sell most of its associated gas to the United States.

However, in March 1979 Mexico announced a policy of greater domestic reliance on gas. This was done first, because of the political repercussions from the aborted gas deal with the United States in 1977, and second, in order to reduce domestic consumption and increase exports of oil for which there was adequate world demand at high prices.

As a result, the government of Mexico decided to encourage a massive shift from oil to natural gas, particularly in the industrial and electrical sectors. In the industrial sector, low-cost gas was offered to new industrial projects. But the most pronounced shift was made by the government-owned Federal Electricity Commission (CFE), which produces 95 percent of all Mexican electric power. CFE mandated the conversion of many of its existing oil-fired plants to gas. It also mandated that all new oil or gas-fired power plants be able to use either fuel and that gas be consumed except in cases of emergency.[5] At the end of 1979, Mexico's government estimated that the CFE system consumed 1.1 billion cf/d and projected that CFE would use 1.4 billion cf/d by 1982. In addition, CFE expects to double Mexico's electricity generating capacity by 1988.[6] If the CFE's fuel mix remains constant between 1982 and 1988, gas demand by the power plant sector will double to 2.8 billion cf/d by 1988.

Estimating future growth in gas demand in the power plant sector as well as the other end-use sectors is necessary in determining export potential because only after domestic demand for gas as well as oil is satisfied can there by any exports.

Domestic demand for oil, particularly refined products, is increasing at high rates mainly because of increasing population, the national industrialization program, and low domestic product prices. Between 1978 and 1979, in terms of sales volume, gasoline demand jumped more than all other major products at 15.8 percent.[7] Kerosenes and diesel followed with increases of 11.4 and 7.8 percent respectively. The average prices for gasoline and automotive diesel fuel are 56 and 14 cents per gallon respectively. Mexico's government has considered raising oil product prices, but that would contradict its traditional mandate of providing low-cost energy to its people. Contributing to the domestic demand is the effect of low prices along the U.S. border. It would be almost impossible to determine how much Mexican gasoline Americans use, but refinery output of unleaded gasoline went up 63 percent in 1979.

EXPORT POTENTIAL

The new gas demand policy means that potential exportable gas from Mexico would only rise to 2.2 billion cf/d in 1988, about half the CRS original estimate. Greater internal consumption of gas should, however, mean more oil for export, about 200,000 to 300,000 b/d by 1988.

It should be emphasized that Mexico does not have to export the estimated surplus gas. Mexico has two viable options to deal with the gas: to "shut in" nonassociated gas production (about one-third of current production or about 1 billion cf/d) and just use gas produced with oil, or to further increase domestic gas consumption. Shutting in nonassociated gas wells can be done easily; the main cost is to the Mexican economy from the loss of gas sale revenues.

The other option is increasing domestic demand even more than already estimated. In the extreme case of Mexico's maximizing a gas substitution strategy, it is estimated that the resulting demand could reduce substantially the gas available for export. Because the government owns and controls CFE and PEMEX, Mexico has great flexibility in implementing a gas demand policy that will mean either a high or moderate domestic demand for natural gas.

Government control over consumption, coupled with the "shut-in" capability in dry gas production, means that Mexico could virtually eliminate gas exports of any significance during the 1980s.

Mexico has raised its oil production platform for three basic reasons: first, the oil finds in the Campeche area have been more impressive than originally estimated; second, domestic demand for oil products like gasoline is increasing at unprecedented rates; and third, there were overcommitments of oil in international trade deals. Oil exports totaled 900,000 in September 1980 and President Portillo stressed that Mexico will meet all of its export commitments, estimated to total about 1.5 million b/d by the second half of 1981. Some of this increase in exports could go to the United States, and the proportion of Mexican oil exports going to the United States could remain constant or drop, depending on how Mexico views its relationship with the United States.

THE U.S.-MEXICAN OIL AND GAS RELATIONSHIP

The U.S.-Mexican oil and gas relationship can be viewed in typical consumer/producer terms. The three basic concerns of oil-importing consumer nations are that they have a continuous supply of oil, in adequate volumes, and at reasonable prices. The producer cares less about the meeting of incremental world demand, and more about: first, avoiding resource waste and technical problems inherent in depleting the resource too fast; second, conserving a resource that is finite; and third, satisfying the economic and political policies made by the producer's government. Thus, in a world oil shortage, there is an inevitable conflict between the consumer wanting more oil, and the producer not producing all the consumer wants, but probably making more money because the price continues to increase.

The conflict is even more accentuated in the U.S.-Mexican oil and gas relationship. As the Middle East becomes more unstable, the natural tendency of the world's largest oil importer is, of course, to diversify supplies, looking toward the geographically closer and more reliable sources. The United States in recent times has paid Mexico's asking price for oil, but Mexico continues to diversify its market for economic and geopolitical reasons. One important consolation, however, is that Mexico's oil does go into the world oil pool, satisfying another part of the world oil demand.

In natural gas, the U.S.-Mexican relationship is still in transition. Mexico wanted to sell the United States large quantities of gas, but the United States did not want to pay the asking price, which was tied to world oil prices. The negotiations fell through, but were then reconvened. The result was an agreement that contained a pricing compromise but a far smaller supply of gas (300,000 cf/d) than originally offered (2 billion cf/d). The agrement can really be viewed as more of a political accomplishment, neutralizing the political fallout from the 1977 negotiations, and providing a foundation for future gas export agreements.

THE FUTURE

In the 1980s, Mexico and the United States will continue their respective producer and consumer roles. U.S. dependence on imports may actually increase, mainly due to faltering levels in domestic oil and gas production. Mexico will enhance its role as an oil and gas exporter, and experience the power that goes with it.

After considering the domestic and foreign policy factors, Mexican oil exports should reach a peak of 2.5 million b/d, and gas exports will reach an exportable peak of 2.2 billion cf/d during the 1980s. Some of the reasons are:

- It makes more sense for Mexico to keep extra oil and gas in the ground because of political considerations such as conserving its national patrimony for future generations and because of economic considerations such as the increasing asset value as the real price of oil and gas go up during the 1980s;
- Less capital will be available for oil and gas development because it will be diverted to more labor-intensive industries; and
- Faster development of the oil and gas industry would create unacceptable inflation rates in Mexico.

A 2.5 million b/d estimate will mean that Mexican oil will account for about 7 percent of the total oil available in world trade in the 1980s. Mexico will use some of those exports to further enhance its status in the third world, especially in Latin America. Mexico and Venezuela together represent a growing political force in the Western Hemisphere with which the United States will have to contend.

Mexico is presently unable to provide "surge capacity" to other nations simply because capacity is already being expanded as fast as possible. All of this capacity must be put into production if Mexico is to reach its 1982 oil production goal. Once the production goal is reached, Mexico might be technically able to develop surge capacity. There are a host of economic and political questions that are attached to such development, and it is impossible to tell whether or under what circumstances the Mexican government would pursue such a course.

Mexico has generally followed the pricing policies of other producers, raising the price gradually for oil purchased under contract. Mexico probably will not hike prices very sharply, unless other producers do.

The United States will probably remain a preferred customer, but less so than now. The United States will probably receive all of Mexico's gas exports and about 50 percent of its oil exports, but there is no guarantee. It will be up to the United States to find incentives such as technology transfer and trade in food and industrial goods to make this supply dependable. Under federal government sponsorship, California and California-based companies, with their technology industries and geographical proximity, can participate in trade deals with Mexico, thereby meeting head on the competition from foreign sources for Mexico's oil. It must be realized, however, that the United States will not have a special right to Mexico's oil and gas, but only a special right to compete for it.

NOTES

1. Philip R. Woodside, Mexican Oil Potential (Washington, D.C.: USGS, February 25, 1980).

2. David Ronfeldt et al., Mexico's Petroleum and U.S. Policy: Implications for the 1980s (Santa Monica: Rand Corp., June 1980).

3. U.S. Library of Congress, Congressional Research Service, Mexico's Oil and Gas Policy: An Analysis. Published as a committee print by the Joint Economic Committee and the Senate Foreign Relations Committee (Washington, D.C.: Government Printing Office, 1979).

4. CRS qualified its original estimate of gas exports because the prolific oil areas in Reforma have large deposits of associated gas that would have to be produced with the oil.

5. U.S., State Department, Industrial Outlook Report: Power Equipment (Airgram, October 1, 1979), p. 7.

6. Antonio Garza Morales, "Mexico Will Have the Capacity to Generate 20 Million Kilowatts of Power by 1982," Mexico City Excelcior, September 3, 1979.

7. "Refining Plays a Big Role in Mexican Oil Boom," Oil and Gas Journal, August 25, 1980, p. 66.

8

POLITICS OF MEXICAN OIL

David Ronfeldt and Arturo Gandara

INTRODUCTION

Reports of vast petroleum resources in Mexico have led to
popular speculation that Mexico could, perhaps should, help to rescue
the United States from its energy security problems. This kind of
popular thesis or speculation has been accompanied by several assumptions: first, Mexico's petroleum is vital to meeting U.S. energy
security needs; second, the United States can strongly influence what
Mexico does with its petroleum; and third, petroleum should be the
crux or heart of U.S.-Mexican relations.

Recent work for the Department of Energy (DOE) has cast doubt
on the wisdom of this conclusion, projecting that Mexico will probably
choose what we call low-to-moderate levels of petroleum production
and exports. Encouraging Mexico to maximize its oil production would
not be in the best interests of either country. U.S. interests in Mexico
are too complex and multidimensional to be dominated by petroleum
issues.

The first section that follows presents a picture of Mexico, discussing what is in the ground, the resources, and what is in the sky,
some ideals of Mexican nationalism. The chapter will then fill in the
foreground with some pragmatic aspects of political institutions and
energy policymaking in Mexico. Then the last part will summarize
some of the findings on what policy outcomes may be preferable for
U.S. interests.

In terms of what is in the ground, there is a 90 percent probability that Mexico will ultimately produce more than 70 billion barrels,
a 50 percent probability that it will ultimately produce more than 90
billion barrels, and a 10 percent probability that ultimate production

will ever exceed 120 billion barrels. That higher level assumes that the fields in the southeast as well as some new areas turn out to be extremely productive. On the basis of this resource assessment, Richard Nehring drew several long-range petroleum production profiles. He concluded that Mexico could become one of the world's leading petroleum exporters—exceeding Iran's former peak of 6 million barrels per day (b/d). It is extremely unlikely that Mexico could ever exceed 7.5 million b/d, and almost totally doubtful that it could ever achieve 10 million b/d as some have speculated. However, if Mexico decides to produce at these higher levels, it will be adopting a profile that could be sustained for only 15 to 20 years.

In the light of Mexico's growing energy consumption at home, Mexico's only rational choice seems to be between a low production profile of less than 3.5 million b/d or a moderate production profile of between 3.5 and 5.5 b/d—achieved sometime in the mid-80s and sustained through the 1990s.

Now, jumping into the sky, the policy discussions in the United States have frequently overlooked that petroleum has more than energy or economic significance for Mexico. It is easy for us as Americans to assume that nothing would be more natural than for PEMEX to become a major petroleum exporter, generating vast revenues for Mexico's development, so that its government and private sectors could move rapidly to resolve Mexico's critical domestic problems. However, such American assessments of the "empirical realities" of Mexico's situation do not take into account that through Mexican eyes, petroleum and PEMEX represent "symbolic realities" of extraordinary, almost mystical significance for the essences of Mexican nationalists. These symbolic realities link petroleum and PEMEX to broader national development concerns in ways that constrain the policy choices open to Mexico's government. Overcoming the symbolic challenges and constraints will probably prove much more difficult for PEMEX than will coping with the practical tasks of exploration and development. The value dimensions of Mexican nationalism strengthen our reasons for believing that Mexico's real choice is between a low and a moderate production profile.

The meaning of petroleum for Mexican nationalism extends back into the nineteenth—and even into the eighteenth—century, but we will just talk about its meaning since the nationalization and expropriation of the foreign oil companies in 1938. Since that time, petroleum and PEMEX have embodied the essences of Mexican nationalism—specifically, national dignity, economic independence, and state sovereignty, which can be called the "fateful triad" of Mexican nationalism. What happens to petroleum and PEMEX goes to the very core of what being Mexican means, or is supposed to mean, for Mexico's leaders as well as the man in the street.

In terms of national dignity, PEMEX is a symbol of pride that Mexico and Mexicans can, through a process of struggling and suffering, prove to themselves and to the world, and especially to the United States, that Mexico can achieve success despite widespread expectations of failure.

In terms of economic independence, PEMEX, whose original goals were to provide energy self-sufficiency and to promote national development through subsidized energy prices, is frequently hailed as the key instrument, the main pivot, the master lever of Mexico's overall economic development. Finally, the creation of PEMEX as a political institution is central to the establishment of respect for state sovereignty within Mexico as well as without.

These three concepts, this "fateful triad," provide the standards by which PEMEX's policies will be, and must be, judged and justified within Mexico, In domestic affairs, they frequently take the form of political rhetoric. But when petroleum and PEMEX become foreign policy issues, the terms of the dialogue change dramatically, and the idealized conceptions of petroleum and PEMEX assume a special, compelling force whose ultimate impact is to constrain the choices open to Mexico's government.

As a result, what often seem to be technical issues to Americans, such as the price of a product or the terms of a contract, may be interpreted by Mexicans as issues of profound political sensitivity for their nationalism. Past negotiating encounters, including the recent gas negotiations, show in particular that Mexico's policymaking elites ascribe high priority to defending state sovereignty. They continue to believe that the state is weak, at a critical stage of its development, and still vulnerable to United States and other foreign pressures. Thus, the growth of petroleum relations with the United States, and other issues involving the United States, will frequently be interpreted more in terms of the risks for Mexico's sovereignty than in terms of the possible economic and technological benefits, which is where Americans tend to put the emphasis.

The significance of these symbolic ideals for petroleum policy is, first of all, that the United States should not plan for Mexico to achieve high production or export profiles during the 1980s. One can even speculate that these symbolic ideals imply a kind of limit or a criterion whereby petroleum exports should never exceed 49 percent of total production, with 51 percent reserved for domestic consumption. Indeed, that implicit criterion has been reflected in some policymaking patterns.

Another point is that the nationalist principles require diversifying petroleum exports away from the United States and toward other markets, even though this diversification will be constrained by the inherent logic of dealing with the United States. A further point is that the nationalist principles will keep Mexico from joining OPEC because that could limit Mexico's freedom of action.

Improvements in U.S.-Mexican relations will require, then, that U.S. policy show increasing sensitivity to Mexican nationalist principles. In particular, U.S. policy actions may have to be more sensitive and responsive to Mexico's sense of sovereignty than to what Americans traditionally consider economic rationality. But Mexico is finding new challenges to modify and transform its traditional nationalism, and adapt it to new principles of Mexican internationalism. The outcome of this tension between nationalism and internationalism during the 1980s will have a profound effect on Mexico's receptivity to ideas for interdependence, community, and partnership with the United States in all issue areas.

THE POLITICS OF MEXICAN ENERGY PLANNING

An original research objective in this area was to describe the organizational relationship of PEMEX to other agencies involved in hydrocarbon policy planning. What emerged, however, was a very interesting picture of energy decision-making authority in transition. This is interesting because it was not expected in a political culture dominated by personal decision making rather than institutional decision making. It was also interesting because of the direction to which the changes point.

Interesting as it may be, of what relevance is it to energy planning in the United States? First, energy policymakers should be aware of the decision-making structure within Mexico and the shifts in Mexican thinking. In addition, the Mexican government is unusually sensitive to approaches, statements, or categorizations. For example, when the Japanese government reorganized its foreign ministry, it put Mexico in the Central American desk, with most of the developing Central American countries. Mexico protested vigorously. Japan then responded by putting Mexico in the North American desk. The Mexican protest was stronger. The solution was to put Mexico in a Latin American bureau, and that was much more acceptable. This type of acute Mexican sensitivity to words and actions makes it important that policymakers in the United States be aware of the issues.

Second, and probably more important, is that states such as California and the United States in general and Mexico may, in the short term, be pursuing divergent energy paths. Therefore, decision makers in the United States should more clearly identify the various currents in Mexican energy planning before embarking on cooperative ventures whose basis for cooperation may be subject to change.

At this time it is important to focus on PEMEX and its relation-

ship to other energy-related government agencies. PEMEX has a
board of directors composed of eleven representatives. Five of them
represent the petroleum workers union and six represent the state.
Two of the state directors represent the Ministry of Patrimony and
Industrial Development, one represents the Ministry of Public Finance,
one represents the Ministry of Commerce, one represents the State
Central Bank, and one represents the national electric utility. Two
facts should be noted about the board of directors. First, although
there is union input at the highest levels, board decisions require all
of the votes of the state. Second, the directors are all presidential
appointments. The president can, in fact, appoint PEMEX management as far down as the subdirectors. This can lead to the conclusion
that, indeed, personal decision making is still dominant. On the other
hand, the institutions are developing into counterweights to personal-
ized decision making. The reasons are several. First, to the traditional opposition to PEMEX, there is now added concern over the
goals of national development. Those who would make social development a priority would subordinate PEMEX decision making to that of
other institutions. Second, there is the traditional fight among institutions anywhere for institutional supremacy. Last, and very important, is the development of technocratic loyalty—a loyalty to a profession or to a particular technology.

To gain an appreciation of the institutional politics, it would be
useful to review the factors constraining petroleum and energy planning. First, the autonomy of PEMEX is being challenged by new institutions such as the Ministry of Patrimony, the Energy Commission,
the Institute for Nuclear Energy, and various other energy-related
institutions. Although the bureaucratic differences are not as visible
as in the United States, the differences are significant. The apparent
common goal of these other institutions is the subordination of petroleum policy to broader energy and economic policies.

Another factor is the high growth of demand for energy within
Mexico. Domestic consumption has to be satisfied first. Mexico has
had very high growth rates of energy consumption—historically, about
8 percent a year. Some sectors will grow as fast as 10 percent annually. There are, in addition, incredibly large increases planned in
electrical production. Between 1960 and 1976, there was a 10 percent
annual growth in electricity demand, and the National Industrial Plan
calls for future growth of 14 percent a year. The electrical sector is
shifting from being the third largest consumer of primary energy
resources to the largest. This is occurring while real prices of
domestic oil, gas, and electricity are declining.

What is the role of alternative energy sources in Mexico with
this kind of expected electrical growth rate? Mexico's future energy
sources are diverse, but most nonpetroleum sources will make a

limited contribution. There are, however, political alliances forming behind new energy institutions. The nuclear option, in particular, is in ascendance because of a combination of nationalism and the strength of the nuclear establishment. These supporters of nuclear energy can be called "nuclear nationalists."

Given some of these factors, what then are the relevant energy institutions and what are their roles? First and foremost is the Ministry of Patrimony. It has the constitutional responsibility over non-renewable resources under Article 27 of the constitution. In 1976, there was a very important administrative law change that expanded the jurisdiction of this ministry, merging it with the Ministry of Industrial Development. The new Ministry of Patrimony and Industrial Development has greater control over relevant state and mixed industries, including PEMEX and the Federal Electricity Commission (CFE). (Mixed industries are those in which the government owns some percentage but in which there is also private investment.) The new ministry also has greater control over foreign industrial investment and technology transfer. In addition to having this important jurisdiction over areas that would affect energy policy planning, the Ministry of Patrimony also oversees the Mexican Energy Commission.

The Mexican Energy Commission was created in 1973. Its role, at the time, was a bit uncertain, and it has had very little influence over petroleum policy. With all the historical and institutional structures built around petroleum, it would be very difficult for a new institution to have an input into that sector. But PEMEX or the petroleum sector is not really reaching out for any kind of jurisdiction in the development of alternative energy sources, so new institutions could build alliances with promoters of alternative energy sources. Working against this possibility is that the director of the Mexican Energy Commission is a nuclear engineer. In addition, in Mexico, as in other developing countries, there is great suspicion of the motives behind the industrialized countries' promotion of alternative energy sources. Now that Mexico has oil, and could vault into the industrial world, why is the United States promoting appropriate technologies? The concern is that these technologies will only save the oil for consumption by the more developed countries, such as the United States. Mexico does not necessarily feel that it should adopt alternative technologies in order to make oil available for the United States. This can be characterized as an influential position within the Mexican Energy Commission. Geothermal energy, whose development is much further along, is a different matter.

As to electricity, its generation in Mexico is nationalized. Like PEMEX, CFE was also created by President Cardenas in 1937, but the political parallel ends there. Nevertheless, we have to ask whether the electrical generating sector is a counterweight to PEMEX.

CFE is in a strategic intermediate position, being a consumer and also a generator of a competitive energy source. Article 27 gives the government jurisdiction over its national resources, but electricity was exempted until 1960. In 1960 the constitution was amended to declare electricity generation, distribution, and transmission in the public interest, paving the way for nationalization. Up to that time, the state-owned electric utility coexisted with foreign private utilities but did not compete with them. It was principally a wholesaler of hydroelectric power. For example, in 1950, CFE produced 50 percent of the power in Mexico but distributed only 20 percent, selling the rest to private utilities, which were generally Canadian or American. In 1960, however, this constitutional change permitted the slow nationalization of the electrical industry in Mexico. So, from 1960 to 1978, CFE's efforts were largely directed at the nationalization or consolidation of electrical production in Mexico. Therefore, it really could not play much of a role in some of the energy-planning sectors.

Why has CFE continued to play a secondary role? First of all, its financial situation was very bad. With 10 percent growth per year, it had to make a lot of capital investments. It had to borrow a lot of money, and around 1975 or 1976, 38 percent of its revenues were directed solely at paying the interest on the debt that it owed. There was another problem: interconnections in Mexico were poor because there were many different utilities that had not been centralized or nationalized. The reserve margin in 1976, for example, was 27 percent. Meanwhile, CFE was continuing to grow, partly because electricity rates were subsidized. This led to governmental review of the electrical sector. In 1977 a commission established some new goals. It said that CFE should plan on a 10.1 percent growth rate and reduce its reserve margin to 15 percent. The government then took over half of CFE's debt and, in return, CFE agreed to these particular governmental goals.

But CFE's plans were disrupted when the Ministry of Patrimony issued its National Industrial Plan, which specified a 14 percent demand growth. CFE had originally cancelled plans for some plants, but now with the 14 percent growth rate, it had to plan on a tremendous capital investment. For example, by the year 2000, there will be 90 new electrical generating plants in Mexico. It has been strongly suggested that 9 will be nuclear, 16 will be oil and gas, 42 will be hydroelectric, 15 will be geothermal, and 8 will be coal.

Since CFE is a principal consumer of energy resources, it is critical to the diversification of energy resource consumption (for example, the utilization of formerly flared gas). Therefore, CFE now occupies a more strategic position in planning. CFE will be more influential in its effect on the consumption of primary resources than is indicated by end-use figures since the conversion of primary resources to electricity is only about 33 percent efficient.

How then does this intersect with the nuclear energy sector and the rest of the energy planning sector? Mexico's Nuclear Energy Commission was created shortly after Eisenhower's Atoms for Peace Program in 1956, but for 16 years it was rather dormant. It had a low budget, and it really did not have a large role. But while it concentrated on isotope generation and medical research, it was laying a very important foundation. The people that it trained—the engineers, the union—constituted a very strong, articulate group for an important new technology.

In 1972, the commission was changed to a more formal institute. Two very important events had occurred earlier that might have precipitated the 1972 change. First, in 1969 the CFE contracted with General Electric (GE) for a light water reactor that would utilize an enriched uranium fuel cycle. The nuclear energy community did not have much of an input into that particular decision. Second, in 1970 there was a discovery in Mexico of 1,800 tons of proven uranium reserves. Since then the increase in uranium reserves has almost paralleled the increase in oil reserves. For example, now there are 10,000 metric tons of proven uranium reserves, 100,000 to 200,000 tons of probable reserves, and 500,000 tons of potential reserves.

The purchase of the GE reactor was a negative but unifying factor. The nuclear community awoke and said, "We are being left out of this country's electrical generating discussion." This may be somewhat overstated, but the community did not have the influence it probably should have had. The uranium discovery, however, provided a much needed boost. For the first time, there was a clear role as to what the nuclear community should do. At the same time, the Canadian reactor option became far more attractive and a rallying point for nationalists and for people who were interested in energy independence. For some, it was clear what Mexico ought to do with the newly discovered uranium: as with oil, it should be used to set Mexico down the road to self-sufficiency.

There were, however, several things standing in the way of the nuclear nationalists. If you look at the historical pattern of the development of nuclear energy, the first few reactor decisions determine what kind of fuel cycle a country will ultimately decide on. In addition, the enriched uranium Laguna Verde reactor is costing more than expected. The original estimate was 4 billion pesos. It is now estimated to cost 23 billion pesos. The enriched uranium price also went up from $7 per pound to $35 per pound. This led to a debate whether Mexico should use its uranium resources to generate revenues to support either the nuclear sector or other government sectors. Consequently, there was a proposal to restructure the nuclear industry by creating ENAN, which would be a research institute, and URAMEX, a counterpart to PEMEX, that would develop, produce, and perhaps export uranium.

There was, however, tremendous nationalist opposition to this particular plan. It was interpreted as a move that would allow private concessionaires to waste the uranium reserves. It paralleled the discussion on oil. Mexico would be squandering its national patrimony; it would be paving the way for foreign investment in the uranium sector; it could not exploit its own technology, perhaps; and it would be an invitation to the kind of foreign influences Mexico had experienced in the past.

So something very interesting happened. There was a modification of the legislation that originally came to Congress. In a hierarchical, top-down decision-making and legislative-producing country like Mexico, it represented an unusual debate and a very significant turn of events. The new legislation that was finally approved still restructured and allowed the separation of the nuclear functions but it gave the nuclear community the assurance that they would be the sole research unit for URAMEX. Second, there were very stringent restrictions on exports, and there were no private concessions allowed. Third, a national Atomic Energy Commission was created to oversee the whole thing.

In summary, there is a vigorous debate over Mexico's future energy policies. Mexico has had, at least in broad terms, a consistent energy policy since 1938, and there appears to be wide consensus that the same policy should continue long into the future. That policy has been one of self-sufficiency or, as it is sometimes put, energy independence. The debate, however, is not over this broad policy, but over the path chosen to implement it. In the past, the self-sufficiency policy has largely been based on a hydrocarbon policy and, more specifically, on a petroleum policy. That policy has been one of investment, exploration, and production by PEMEX in order to satisfy internal demand and to provide petroleum products at subsidized prices in order to stimulate industrial development. A continuation of this basic policy, with the important addition of substantial petroleum exports, would serve the traditional interests, PEMEX, the Institute for Mexican Petroleum, and the subsidized industrial sector. At the same time, such a policy would operate in disregard of new institutions, more specifically, the Ministry of Patrimony and Industrial Development, the Energy Commission, the nuclear power interests, and the nationalist groups who are seeking some structural and distributional reform. The traditional interests are sure that the present path is in harmony with past policies. The reformers are concerned that, unless they succeed in changing the present path, the present structure, unacceptable in their view, will be cemented and the possibilities for reform will substantially diminish.

Both of these interest groups, however, frame their concerns with the same rationale of energy self-sufficiency or independence. Neither group raises the more serious question of whether this is a viable goal. Those that base their independence on a hydrocarbon policy are clearly aware of the nonrenewable aspect of this resource and turn instead to the concept of converting nonrenewable resources into renewable wealth. The other group seeks energy independence through a natural uranium fuel cycle, and perceives that the uranium reserves could indeed be a renewable natural resource through re-processing, eventually creating a plutonium-based energy sector. The irony is whether any greater independence is then achieved. This path would require dependence on foreign technology. Furthermore, while independence from enriched uranium would have been accom-plished, a new dependence on heavy water would have been substituted.

The Mexican concern for energy independence and self-suffi-ciency is reflective of the broader concern for independence and self-sufficiency in all the nation's activities. In both areas, however, this concern cannot serve as more than a diversion from the reality of interdependence, of which all nations, large and small, have become acutely aware. In addition, framing the issue as one of independence falsely suggests the binary states of a nation's relationship with an-other nation, dependent or independent. History teaches us that these relationships lie along a spectrum. It is far less clear whether the threshold between independence and dependence has been crossed as the political and economic activities of two nations become more inter-twined—whether by choice or circumstances.

CONCLUSION

So one ends up with a picture of Mexico in which petroleum poli-cies are supposed to be subordinated to energy policies, these to national development policies, and all of this to the ideals or the tra-ditions of Mexican nationalism. At the same time Mexico's political and economic institutions are going through a major period of trans-formation. That is why Mexico may be heading toward a "crisis of success."

What are the stakes in all of this for the United States? They can be seen in two areas: energy security and U.S.-Mexican relations. In the area of energy security, U.S. objectives should take into ac-count, first of all, that we do not have an interest in Mexico exhausting its resources before the turn of the century. We also do not have an interest in seeing a relatively secure energy source become an insecure energy source because of the potentially destabilizing impact of high petroleum revenues.

In the area of U.S.-Mexican relations, one must look at three levels. First of all, Mexico is important simply because it is a neighbor. The United States has an interest in having a stable, progressive, and friendly Mexico as a neighbor. Mexico is now at a critical stage of its institutional development. What happens in this stage is going to be affected by relations with the United States regarding not only petroleum but also immigration, trade, and other areas.

The second dimension concerns the massive social, cultural, and economic interconnections that are growing between the two countries, particularly at the border level, but spreading deep into both societies. Our two countries are becoming inextricably entangled. This is most noticeable at the border level, which could become the melting pot or the boiling pot of future bilateral relations.

The third dimension is that Mexico is an emerging medium power, which is now important within the framework of U.S. global interests—not just because of petroleum, but also because Mexico is our fourth most important trade and investment partner and likely to become second only to Canada by the mid-1980s. Furthermore, Mexico is showing new determination and new ability to influence the formulation of international doctrines and multilateral principles, and not only in energy and trade. Also, in the future, Mexico may be important to security within the Caribbean basin.

From this view of U.S. interests are derived some preferred policy objectives in five areas. The first is the production export level and rate of development. The central issue in Mexico's petroleum development, and the issue of greatest concern to the United States, is Mexico's choice of overall production levels for oil. It can be concluded, from looking at the profiles mentioned earlier, that the choice of a moderate level would be best for overall U.S. interests. The moderate level would mean 5.5 or 6 million b/d with exports ranging between 2.5 and 1 million b/d in the late 1980s through the 1990s, depending on Mexico's domestic consumption patterns. The choice of a low production profile, less than 3.5 b/d, would do little for U.S. energy security interests. The choice of a high production profile, over 3.5 or 4 million b/d, could induce economic and political disruption within Mexico. It is too early to state what Mexico's eventual choice will be. Current coalitions appear to favor either moving up very slowly from the current level, thereby committing Mexico to a low production profile, or moving up more rapidly to the moderate level by the mid-1980s. There appears to be no significant support for moving to high production and export levels, in part because of fears that Mexico might become another Iran or Venezuela. We foresee that Mexico's evolving definition of its own interests and objectives will gradually result in the choice of a moderate production profile. Because this choice is consistent with U.S. interests, there appears to be no need to offer Mexico any special incentives or concessions, or to fashion any special pressures designed to get them to move up more rapidly.

A related issue is whether or not Mexico needs additional technical assistance to develop its petroleum resources. PEMEX's technical capabilities are impressive; it is quite capable of moving up to the moderate level at current rates, so this idea of special technical assistance for Mexico's petroleum development is a nonissue.

Another issue area is emergency production capacity. It would be in the best interests of U.S. energy security that Mexico have excess or surge production capacity that could be used to raise exports rapidly during an international emergency or natural disaster. Because development drilling will probably outstrip production ceilings, especially if Mexico decides to keep exports low, there is a strong likelihood that Mexico will have excess capacity by the mid-1980s. Whether or not Mexico installs all of the equipment to make that excess capacity usable is still open to question. If Mexico decides to develop a policy in this area, one of its key objectives will surely be to protect its sovereignty. Mexico would probably not be amenable to any special bilateral arrangements for emergency production with the United States, and its thinking would surely move in terms of multilateral frameworks. In this area there is a possibility, not a large one, but a possibility, that Mexico could explore some kind of special agreement with an existing multilateral agency, the International Energy Agency, for using excess production capacity under emergency terms that Mexico could specify. Yet another possibility would depend upon taking a lesson from President Portillo's speech to the United Nations in 1979, in which he proposed a global energy plan. Mexico might be willing to think about an emergency production capacity arrangement that would involve several suppliers as well as various producers, perhaps not on a global basis but on a regional or subregional basis, for example, involving Venezuela. These are speculations, and as President Portillo noted in his own speech, the first problem is really to try to formulate the problem, without raising fears of complicity, subordination, or touching other political sensitivities.

A third issue area is export destination and composition. Mexico currently plans to diversify its oil export destinations, reducing the United States' proportion from a current level of about 75 percent down to about 60 percent. The United States could think of several alternatives in this area. One would be to do nothing—let international market forces work. Another would be to seek a bilateral arrangement with Mexico to guarantee that we get a high percentage of its crude exports. A third approach might be to make the U.S. market as attractive as possible. We think the third alternative is preferable. For the United States to receive a higher share of Mexican crude has marginal advantages, but it probably would not be worth the policy effort necessary to assure it, particularly if the policy effort would arouse the flames of Mexican nationalism in a counterproductive manner. Instead, PEMEX's unwillingness to offer price discounts, except recently in the Caribbean basin, will probably constrain it from carrying its diversification policies to an extreme. One unilateral step the United States might take to make its markets attractive would be to remove all price controls affecting refinery investment. Future Mexican crude exports will probably emphasize heavier crudes from offshore, and if U.S. refineries are to make the necessary investments to be able to process these heavier crudes, they will need incentives.

A fourth issue area concerns petroleum revenues and trade diversification. This is related to Mexico's desires to diversify its oil exports, which will be used to diversify its pattern of international trade and commerce. It may well be more important for U.S. economic interests that Mexico's oil revenues come up to the United States than that the actual oil comes here. More is at stake than just recycling. The way in which Mexico spends its oil revenues, particularly in the United States, will have a significant effect on other dimensions of our relationship, such as illegal immigration and the prospects for future production sharing in industrial and agricultural areas within the North American region.

A moderate petroleum production level would most enable Mexico to diversify its overall trade patterns. If Mexico keeps petroleum production low, it will have great bargaining leverage but it will not have quite enough revenues to diversify the overall pattern of trade. If Mexico tries to export at high levels, the country will be so swamped with revenues that there will be a strong inertia for most of those revenues just to come to the United States. So, in the near term, moderate export levels would give Mexico the best advantages for diversifying its trade. The country has been trying to work package deals, with the French, the Canadians, the Japanese, and others.

That strategy will work, but only within limits, and the limits will begin to appear, particularly when Mexico's new economic development policies begin to succeed at developing new export industries that need new markets. That is the point, in the mid- to late 1980s, when Mexico will need to come back to the United States. Thus, it is seen that trade issues will probably prove much more controversial and delicate to negotiate in the future compared to energy issues.

Our fifth and final issue area is energy use and petroleum substitution in Mexico. It was noted previously that there was a rapid growth of domestic energy demand. This is tied to the subsidized domestic pricing policies, to the point where a colleague in Mexico commented recently that Mexico's domestic petroleum policies have a more distortionary effect on Mexico's economic growth than do Mexico's export petroleum policies. It would be in U.S. interests for Mexico to begin thinking about energy conservation, both through bringing domestic prices more in line with international prices, and possibly by adopting new energy-efficient technologies. At the current time, the principal alternative to petroleum is nuclear. Now, while we have an interest in seeing Mexico change some of its domestic petroleum policies, there is probably no particular role for the United States in this area. It would smack of interference with Mexico's domestic affairs. Fortunately, policy trends in Mexico indicate a growing awareness of needs in this area. In fact, the U.S. role may be limited to technical assistance.

9

VENEZUELAN OIL

Frank Tugwell

Venezuela is currently the fourth largest producer in the international system and will continue to be a significant petroleum-exporting country into the future. The production prospects for Venezuela are for a stable and gradually declining output into the 1990s, and a gradual increase in the production of processed, unconventional oil beginning in the early 1990s, but no significant production in unconventional oil until the next decade. There are few things that the United States can or should do to influence the decisions the Venezuelans will take about the quantities they produce and about the destinations of their oil.

The stabilization of production levels has some important implications for the Venezuelan political system. In fact, an analogy between Mexico and Venezuela can be drawn. There may be an underestimation of the appetite for income that can grow and develop and become habit-forming in a country like Mexico—a country with a large population of poor people. If a hypothesis were developed here, it would be that to the degree to which the Mexican political system becomes more open, more democratic, and more participatory the temptation will be even stronger to follow a production rate that brings in more income from oil. In Venezuela this is something that can be called fiscal saturation. Never underestimate the capacity of people to consume, especially where there is great inequality of income and where the development process is just getting underway. There is going to be a lot of political pressure in Mexico that derives from these forces in favor of very high production. It is not known whether we can go up an international learning curve in this respect. It may well be that the mistakes that Venezuela has made—mistakes about which the Mexicans are very concerned—can be communicated and will provide the basis for a much more sensible economic development program there.

A few things must be said about the supply prospects from Venezuela. Venzuela is an old producer. It has been producing oil for five to six decades, and the oil from its conventional reservoirs is running out. This is well understood there. Because of this, the government has determined to level off production, which currently is around 2.2 million barrels a day (b/d). Domestic consumption is around 300,000 b/d. One can foresee a leveling off and then a gradual reduction in output in the late 1980s to around 1.7 million b/d. This is partly as a result of the same thing that is going on in the other OPEC countries, that is, increasing domestic consumption based on low domestic prices for energy. The consumption of gasoline in Venezuela is growing at the rate of 11 percent per year, and the government, despite a commitment when it first came to power, has not increased domestic prices.

Venezuela has determined the level of production that it believes is optimum in conventional oil on the basis of a reserve-to-production ratio that it feels will permit the largest amount of long-term recovery of its resource base, and this is really the basic underlying justification for its production control.

In regard to conventional oil, the heavy petroleum substance that is found in a wide inland tar belt in Venezuela is an enormous petroleum resource. The estimates that earlier were floating around claimed that the belt contained 700 billion barrels, of which perhaps 10 percent could be recovered. Today, the estimates range from 1 trillion to 7 trillion barrels, and perhaps an even higher recovery range because of the efforts that have been made to improve the systems for producing and processing the material. That this is a lot of petroleum is suggested by the fact that the current world reserves are somewhere in the range of 650 billion barrels. We are talking about a reservoir greater than this. Venezuela, however, does not view its heavy oil as a product that should be competing with conventional oil in current markets. Heavy oil is usually very heavy and difficult to extract (although it is often extractable by conventional or secondary recovery techniques). But more important, it has significant contaminants in it: sulfur, vanadium, and a number of other undesirable elements. Before it can be used, it has to be processed and the processing can be quite expensive. Thus, Venezuelans view this material as something that will gradually come into the market, as the synthetic liquid substances derived from coal and oil shale begin to be prominent in the market, and as the price goes up. They foresee a situation in which world oil production peaks and begins to decline and prices go up high enough to begin to bring on line the synthetic crudes. It is at this point that Venezuela will begin to produce economically this heavy material. There are long lead times involved in the construction of the plants that are needed to process this material. Their forecasts and the forecasts of others indicate that they will be producing probably 300,000 b/d of heavy oil by the year 2000 and between 100,000 and 200,000 b/d by the year 1990. We therefore can foresee a gradual decline in conventional petroleum and a gradual increase in heavy oil produced from the Ornoco Region. The key to Ornoco production, as far as Venezuelan development is concerned, is that it is so expensive that very little will come to the Treasury, currently or in the near future. If it costs $35 to produce it and the market is $40 a barrel, you are only going to get $5 per barrel. It makes sense to go after secondary and tertiary recovery of conventional oil before you begin to do this. This basically is the attitude of Venezuela.

In regard to Venzuela's general orientation toward the world oil system and about U.S. policy options, it should be noted that Venezuela was one of the early nationalistic oil countries. In fact, a Venezuelan oil minister was very important in the creation and founding of OPEC. Generally speaking, however, Venezuela has been a responsible hawk as far as prices are concerned. It has always wanted higher prices, but at the same time it has had a concern for the possible impact of higher prices not only on the advanced societies, but also on the developing countries. It is likely to continue this position. As a stable producer, price increases will continue to be important in order to allow it to satisfy that appetite for income that has become so ingrained.

Venezuela is now paying more attention than it did originally to the careful development of its petroleum resources and to its industrialization program. Nevertheless, one more doubling of income may well lead to a leveling off in available increments, which in turn will have repercussions in the political system in a decade or two.

What policy options might the United States pursue in order to influence production products? American energy experts and government representatives have made repeated trips to Venezuela, trying to convince Venezuela to increase its production very rapidly from the Ornoco tar belt. Often, representatives had hopes of getting U.S. corporations involved in that development process. Venezuela nationalized its petroleum industry in 1976, and there is absolutely no chance of American corporate equity participation in the development of the tar belt or even of conventional fuels in that country. Technology transfer—the transfer of something like Exxon's flexicoking process to manage the tar belt materials—is possible; but not an equity participation. Venezuela can almost certainly buy the technology that it needs. It can bring technical people in and can develop the skills in the time frame we are talking about to manage its own resources. The United States presently has a bilateral agreement to explore this. Venezuela is also conferring with Canada, and has a close relationship with PetroCanada.

As far as marketing their oil is concerned, the Venezuelans have been moving away from the United States. They are strongly committed to circumventing the U.S. oil companies as marketers of their product. They now market independently about 40 percent of their oil and want to continue to increase this percentage as well as diversify their markets. In the past, they were very closely tied to the American East Coast residual fuel market, but are trying to move away from that. This strategy fits in very closely with the U.S. oil backout legislation, which will in a sense be pushing out Venezuelan oil. There is an irony here because that legislation is designed to increase the security of the United States. What it is likely to do is push out the single most secure international supplier the country has ever had, while leaving intact U.S. dependence on light oil producers from the Middle East. One should not equate the reduction of imports with an increase in national security. It is where those imports are coming from that is most important, and current national strategy for backing out imports may be counterproductive. In designing a backout strategy to increase national security, the United States should not push Venezuelan oil out of the system.

10

OIL IN INDONESIA

Guy Pauker

INTRODUCTION

According to the California Energy Commission, [1] 16 percent
of the total crude oil input to California refineries in March 1980 was
imported, or a total amount of 292,800 barrels per day (b/d). Almost
all of this comes from Indonesia. A small amount was coming from
the Middle East and has been gradually displaced by Alaskan oil.
Therefore, as a foreign supplier, Indonesia is obviously of unusual
importance to California.

Two or three years ago it was often stated that oil-producing
countries had to be divided into two groups: those like Indonesia,
Nigeria, and Iran who, having large populations and ambitious development plans, are forced to produce to capacity and therefore do not
raise problems for the importing countries in terms of the reliability
of their supply capability. They would produce and export to the limit.
The other countries, of which Saudi Arabia, Kuwait, and the Emirates
were the obvious examples, have very small populations. While they
have ambitious development programs, they can with much greater
ease limit their production, for economic or political reasons. But
that dichotomy was spurious. We have by now at least two examples
of countries which, although they have large populations and ambitious development programs, have either curtailed production in the
recent past, as is the case with Iran, or are seriously opposing the
rapid increase of their future oil production, as is the case with
Mexico. The latter has a population of 75 to 80 million people and
very high unemployment, yet there are strong forces at work to limit
the expansion of oil production for the foreseeable future. Although
neither Indonesia nor Nigeria have yet shown any indication that they

might also curtail their production, it is not inconceivable, at least in the abstract, that they might also decide on a different oil policy, for political reasons, as was the case with Iran, or for reasons of resource economics, as is the case with Mexico.

Indonesia is a country that had a population of 145 million in 1980. It constitutes about one-half of the total population of the OPEC group of countries. But it is only producing about 5 percent of the total oil produced by the OPEC countries, an indication that, relatively, Indonesia is a very poor country. Because of the very rapid recent increase in oil prices, and the very strong situation in the commodities market in general—rubber, tin, lumber, and other metals and minerals—Indonesia has a very favorable export situation right now. And yet, all this has only raised the per capita income of the Indonesia population to an average of about $400 per year.

Indonesia is to this day one of the smallest per capita consumers of energy in the developing world. Even that has to be further qualified. The total electricity production in Indonesia is only a few thousand megawatts a year, which is extremely small. But these figures are, of course, mere statistical exercises because in reality one cannot say that each Indonesian consumes 150 or 200 kilowatt hours per annum.

What really happens in Indonesia is that out of the total population of 145 million, perhaps 1 to 1.5 million people live in the age of electricity and may have a consumption comparable to that of Singapore, roughly 1,700 to 2,000 kilowatt hours per annum. The rest of the population does not consume any electricity at all, if you are willing to discount the fact that quite a few now have battery-operated transistor radios. But that is the only contact the overwhelming majority of the population has with electricity.

Most of Indonesia's population relies on what we call "noncommercial" sources of energy. In Indonesia's case, this consists primarily of firewood for cooking. The firewood is collected from the forest and in many instances, especially in Java and Bali, at considerable risk to the environment. Trees are cut down from the watersheds, affecting the whole ecological balance, changing rainfall patterns and increasing erosion. It also accelerates deforestation caused primarily by the need for more agricultural land. As economic development progresses at a 6.5 to 7.5 percent GNP rate of growth per year, more and more people are going to shift slowly from the use of firewood to commercial sources of energy, primarily kerosene. Although kerosene is a much more costly fuel, it is more convenient than firewood, which has to be gathered and carried from increasingly remote places to the villages in which the population lives. Furthermore, as discussed later, the government keeps the price of kerosene low. Therefore, the pressure of increased consumption of commercial sources of energy and, in particular, of kerosene is great and will undoubtedly continue to increase as time goes on.

Indonesia has the image of a resource-rich country. This is not true if one takes into account the size of its population and the amount of resources needed to increase the country's productive capacity. Conservative estimates of Indonesia's proven reserves of oil are on the order of about 10 billion barrels recoverable with present (primary recovery) production techniques. The Indonesian government and the managers of Pertamina, the national Indonesia oil company, claims that Indonesia probably has reserves closer to 50 billion barrels. These figures are not internationally accepted, and are not the figures on which the most prudent analysts in the world at large are willing to operate. And 10 billion barrels of oil at the present rate of production means that Indonesia will probably exhaust its oil reserves before the turn of the century unless substantial new reserves are found or enhanced recovery is possible.

Besides Indonesia, there are only a few countries in the Pacific Basin that have significant oil reserves. Malaysia, a much smaller country than Indonesia, had a population of only 14 million in 1980, has probably about 3 billion barrels of oil and will remain for the next couple of decades a reliable source of exports. The other important oil country in the Pacific Basin is China. The figures on China's oil reserves are speculative. When former Secretary of Energy James Schlesinger came back from his official visit to China, he told a press conference in Tokyo that the Chinese assume that they have about 100 billion barrels of oil in the ground. About half of it is offshore, which is not yet developed at all. The other half is onshore, of which only a fraction is in production today. More conservative estimates are as low as 20 billion barrels. For a country with a population of 1 billion people, that is not much. It is quite evident that as China's industrial development progresses, the likelihood that China can become a substantial exporter of oil is very low. If one looks at the Pacific Basin as a whole, one can perhaps count on 40 billion barrels of oil as more or less proven reserves. Richard Nehring has concluded that the statistical likelihood of finding new supergiant fields anywhere in the world is not very good. Miracles are not likely to ease the oil crisis in the Asian part of the Pacific Basin. Oil is much more abundant in the Western Hemisphere: in Canada, the United States, Mexico, and Venezuela, and to some extent in Ecuador.

INDONESIA'S OIL PRODUCTION OUTLOOK

Although the long-term prospects of the Indonesian oil industry are controversial, the short-term outlook is good. Petroleum Report,

produced by the U.S. embassy in Jakarta, which has maintained for
a number of years very competent economic officers studying the oil
sector, is fairly optimistic. After several years of slack in exploration, investment in exploration is increasing at present. The relationship between the private foreign companies, Pertamina, and the government of Indonesia have improved substantially in the last couple
of years. Therefore, the situation now looks quite a bit more promising than it looked two or three years ago. The June 1980 U.S. embassy
report entitled "Indonesia's Petroleum Sector" stated:

> The medium-term outlook for Indonesia's petroleum sector
> is now the brightest in over five years with production hold-
> ing better than earlier predicted and a strongly based accel-
> eration in exploration providing the hope that Indonesia's
> crude oil production may increase over the next two to
> three years, reversing over two years of gradual decline.

Before the regime headed by President Suharto came into power
in 1967, there were three major oil companies in Indonesia. They had
obtained their concessions in colonial pre-World War II times from
the Dutch, and were, until 1967, the only producers. No new concessions had been made available, with one small and unsuccessful exception in the early 1960s. These three were Caltex, which is a joint
operation of Standard Oil of California and Texaco, Stanvac, and
Royal Dutch Shell. After 1967, the new regime in Indonesia set as its
major policy goal rapid economic development, and adopted very
favorable foreign investment laws and regulations that widely opened
Indonesia to the foreign oil industry. Production increased substantially and peaked in 1977 at 1.7 million b/d. By that time, about 25
foreign companies were active in Indonesia, with 20 American companies producing 85 percent of Indonesia's oil.

But in the meantime, there had been considerable financial
turmoil and upheaval in Pertamina. In 1975, the Indonesian government and Pertamina found it necessary to change the terms of the
contracts they had with the foreign oil companies. The result was a
very drastic decrease in investment in further exploration. Production
started slowing down. In the first half of 1980, it was only 1.6 million
b/d. This was not a very great decrease, but certainly not an increase.
Also the ratio of proven reserves to annual production was decreasing
as the result of reduced exploration. Pertamina became aware of the
dangers of the situation and more favorable production-sharing contracts have been offered to the oil companies in the last two years.
The U.S. embassy estimated that, in 1980, exploratory budgets might
exceed $600 million, compared with only $140 million in 1977, and the
number of exploratory wells might exceed 180, compared with less

than 100 in 1977. Nevertheless, it is not expected by those who follow
the situation closely that Indonesian oil production will increase very
substantially in the foreseeable future, unless statistically unlikely
major finds occur in the years ahead. The REPELITA III, the Third
Five-Year Plan, has set a target of about 1.84 million b/d for 1984.
Therefore, Indonesia might over a number of years, perhaps by the
mid-1980s, increase production by a modest amount, perhaps a few
thousand barrels of oil a day. But it is not likely to be an explosive
increase and it will not change the energy dependence of the region
on the Middle East and the Western Hemisphere.

In Indonesia, a country in which most of the population consumes
very little commercial energy, economic development is bound to
increase domestic demand and therefore reduce the amount of crude
oil available for export. This will take two forms: first, a good many
people, especially in rural areas, who are now relying on noncommercial fuel (i.e., firewood) for their cooking needs, are going to
shift to kerosene; and second, domestic demands for industry, for
generation of electricity, for transportation, and all the other normal
uses of energy in an industrializing society will increase. During the
past seven or eight years, demand for petroleum products has increased by over 12 percent annually. At present, domestic consumption is 400,000 b/d of crude oil. Extrapolating from these figures,
domestic demand could absorb all of Indonesia's production by the
1990s, leaving no petroleum for export.

In the summer of 1977, the first year of the Carter administration, the White House became interested in helping third world
countries develop alternative sources of energy for three reasons.
First, their demand for world oil supplies would be reduced. Second,
the urge to develop nuclear power as an alternative source of energy
would be contained, with significant benefits to world peace by reducing opportunities for proliferation of nuclear weapons. Third, alternative, decentralized sources of energy, fitting more easily into the
rural environment, would satisfy one of the basic concerns of American foreign economic aid policy, namely, to respond to basic human
needs in the developing world. The result of this policy interest was
a flurry of activity financed and sponsored both by the Department of
Energy and by the Agency for International Development to help countries of the third world develop alternative sources of energy.

Simultaneously, countries like Indonesia independently reached
the conclusion that they should develop alternative energy sources. A
document produced in the Indonesian Department of Energy in May
1980 discusses in some detail what Indonesian energy policy is at
present, pointing out that they first started thinking about an energy
policy in 1977. At that time one of the most knowledgeable and thought-
ful managers of the Indonesian petroleum industry, Ir Wijarso, was

presenting to professional audiences in Jakarta a "doomsday scenario." The scenario pointed out that at the current rate of energy demand growth, Indonesia would run out of oil for the export market before the end of the century, and might exhaust its oil reserves altogether. Indonesia decided to cope with the problem by developing alternative sources of energy. The U.S. government then came into the picture and focused on concrete programs that were less familiar to the Indonesians, such as solar energy. Public awareness of these issues in policy circles has grown very rapidly. It is now established policy that Indonesia must increasingly conserve oil for export, for domestic petrochemical industries, and for those uses for which there are no substitutes, primarily transportation. Other energy requirements must be satisfied by other available energy sources, namely natural gas, coal, hydroelectric power, geothermal power, and perhaps nuclear energy.

Indonesian Department of Energy officials believe that they are making some progress in that respect and hope to make further progress, as their five-year plans develop. But they are not excessively optimistic about what this will really amount to in the foreseeable future. Statistics produced by the Indonesian Department of Energy show that the share of oil which provided about 82.2 percent of total energy consumption (24.8 million tons of coal equivalent) in 1978-79, can be decreased by 1983-84 to only 77.7 percent (which would then be 38 million tons of coal equivalent), while all nonoil sources of domestic energy consumption would be increased from 17.8 percent to 22.3 percent. This would take primarily the form of substantially increasing the use of natural gas from 15.8 percent to 17.6 percent (from 4.7 million tons of coal equivalent to 8.5 million tons of coal equivalent). Hydroelectric power will be increased from 1.3 percent of the total to 2.2 percent (from 398,000 tons of coal equivalent to 1 million tons of coal equivalent), the coal from 0.7 percent to 2.5 percent (from 209,000 tons to 1.2 million tons).

Even these expectations are probably optimistic, because Indonesia is a country that is still in the relatively early stages of economic development. Its managerial capacity is limited to a very few, very able people at the top. These top people have to cope with a rather inexperienced and not very efficient bureaucracy. So translating these policies into effective programs will create real problems. However, if their estimates are even partly fulfilled, that will be a valuable contribution to what they very lucidly see as the need to conserve oil for those utilizations for which there are no easy substitutes, transportation and petrochemicals, and to reserve as much of their oil as possible for export, which today is Indonesia's number one source of revenue.

How efficiently they will be able to develop alternative sources of energy and save oil for the purposes for which the planners and policymakers would like to use it, is still an open question. The pressure not only to make more kerosene available to the population, but also to keep kerosene at a price that will be acceptable to the population and therefore avoid what could be very serious political consequences, is not an easy one to withstand.

Last year the Indonesian government was undergoing a very agonizing internal policy debate on whether the price of kerosene should be maintained at a very low level. Those in charge of internal security management favored keeping the price low. But the economic planners of the country, the Department of Finance, and BAPPENAS (the National Planning Agency) were arguing that that price had to be increased because it was just unrealistically low. Finally, after considerable agonizing, the Coordinating Minister for Economic Affairs and head of BAPPENAS, Widjojo Nitisastro, took almost one hour of live radio time on April 30, 1980, to announce and explain the government's decision to raise domestic fuel prices by 50 percent.

For domestic reasons, the price of kerosene per liter was increased only to 37.50 rupiahs, which is $.06. They are trying to compensate through cross subsidization, by increasing the price of gasoline and of other petroleum products much more substantially. The price of super gasoline was increased to a total of 220 rupiahs per liter, about $.33 per liter. These are very serious policy issues in a country that has to deal with a very poor population. What these subsidies in the domestic petroleum products consumption sector represent is illustrated by the following figures.

For fiscal year 1979-80, according to the World Bank, the oil subsidy totalled $960 million. Despite the price increase, it is expected that the oil subsidy will cost the Indonesian government $1.28 billion in 1980-81. Even assuming further price increases of about 25 percent during 1981-83, the World Bank, nevertheless, concluded that by 1983 the domestic price of petroleum products would still be only half of the economic cost, and added:

> A realistic pricing policy as part of a comprehensive energy policy package should result in a deceleration of domestic oil consumption growth. Our projections assume that the growth rate would nevertheless average about 10 percent a year up to 1985, compared with 13.5 percent during the years 1973-79.

As for developing alternative energy sources, in practice it boils down to what can be done with natural gas, coal, hydroelectric power, geothermal power, and nuclear power.

DIVERSITY OF ALTERNATIVE ENERGY RESOURCES

Indonesia is a country that has reserves of natural gas. Again, the amount depends a great deal on semantics. Reserves are of the order of 30 trillion cubic feet. Thirty trillion cubic feet is about the consumption of the United States for 18 months. Therefore, in the total Pacific Basin picture, this is not a very large amount.

From an internal point of view, Indonesia faces much the same problems with natural gas as with crude oil. It is a major source of foreign exchange as an export product in the form of liquefied natural gas (LNG). It is not very suitable for domestic consumption, since most of it is located far from populated areas.[2] The population is mostly in Java and Bali, which together contain about two-thirds of the total population of the country, whereas the natural gas reserves are at Arun in North Sumatra, which is a thinly settled area, and at Bontang and East Kalimantan. Therefore, this gas is more readily available for liquefaction and transportation by cryogenic tankers to importing countries such as Japan.

To use the gas domestically, Indonesia would have to build LNG facilities, pipelines, or petrochemical plants, all of which would be very expensive. Cryogenic LNG facilities are expensive anywhere. Pipelines would have to transverse a country that spans about 3,000 miles from one end of the island chain to the other. The distance between the northern top of Sumatra and Jakarta exceeds 1,200 miles. It is probably not feasible at present to build gas lines under the Java Sea, so deliveries of gas from East Kalimantan are impossible. Most important, Indonesia could not command as high a price in the domestic as in the export market, making it difficult to pay for the massive investments.

By exporting LNG from Arun and Bontang, Indonesia expects to earn $6 billion annually in foreign exchange during the 1990s. Some of it may come from the yet undeveloped field in the Natuna Sea, which may raise jurisdictional conflicts with China and Vietnam.

From the point of view of a foreign investor concerned about the security of supply of petroleum products, investment in LNG is more attractive than investment in oil. Once produced, oil can easily be shifted from one consumer to another by ordering tankers to a new destination. During the 1973–74 oil crisis, tankers were diverted quite regularly from one place to another, just by one telex message. But LNG involves a very expensive and complex infrastructure at both ends as well as appropriate cryogenic tankers, in many cases very specifically tailored for one particular operation.

California and the United States have fallen behind Japan in assuring itself part of the natural gas supplies of Indonesia. The Japanese are very actively developing this LNG resource. Japanese

utilities are signing new contracts to build additional trains, both in
North Sumatra and in East Kalimantan. The Indonesians have shown
considerable patience and goodwill in waiting for U.S. federal and
California state authorities to complete their lengthy and cumbersome
decision-making process. Their patience may be due partly to friendly
feelings toward the United States and partly to shrewd business sense.
That is, it is better not to be exclusively dependent on one country,
Japan, for all LNG exports. In any event, California consumers will
suffer if a 20-year supply of Indonesian LNG is lost because of bureaucratic delays.

Indonesia also claims to have perhaps as much as 18 billion tons
of coal, with 4,000 to perhaps 7,000 kilocalories (kcal) per kilogram
(kg) calorific content. Most of this coal is located in Central and South
Sumatra and East Kalimantan. In colonial times, coal represented a
fairly substantial part of the total commercial energy consumption of
the country, but as oil took over, production declined from over 1
million tons before World War II to 200,000 tons in 1978. The coal
mines were neglected and became a negligible factor in the total energy
picture; railroads were electrified or dieselized and coal ceased to be
used. In the last few years, as part of this recent effort to save petroleum for its optimal uses, there has been a lot of talk in energy policymaking circles in Indonesia about using coal in at least two major
ways: first, for electricity generation; and second, as a substitute
for kerosene for the domestic household market, which has apparently
been very successful in South Korea. The first of those two potential
uses seems more promising than the second. There are now active
plans under way and tenders have been issued for preliminary engineering work to redevelop some of the South Sumatra coal mines. At
present prices, Sumatran coal could not compete in the export market
with Australian coal. But according to the Indonesian Department of
Energy, Sumatran lignite could be shipped to the western tip of the
island of Java to fire about 800 Megawatts of Electricity (MWe) of
coal-burning power plants that would supply the future power grid of
the overpopulated island of Java, which must industrialize in order
to survive. Even after rehabilitating the Sumatran coal mines, building
a railroad, necessary port facilities, and the power plants, the project
would be cheaper than a nuclear power plant of the same capacity.
Furthermore, Indonesia could avoid becoming dependent on foreign
uranium. About the environmental impact of coal-burning power plants
on the west coast of Java, there have not been any evident studies or
comments in Jakarta.

Another source of energy that is actively discussed and modestly
developed in Indonesia is geothermal power in West and Central Java.
The step from pilot projects of a few tenths of megawatts to something
that would make a real contribution to the total energy supply is a big

one that has not been taken. However, geologists estimate Java's geothermal potential at 5,500 MWe, Sumatra's at 1,100 MWe, Sulawesi's at 1,400 MWe, and perhaps another 2,000 MWe in other islands, except Kalimantan.

Hydroelectric power has been developed to some extent. In Java, about 200 kilometers from Jakarta, the Jatiluhur project, yielding 120 MWe, was underused for a number of years in the early 1970s. Transmission facilities were not adequately developed and the population lacked purchasing power. In the capital city of Jakarta, according to reliable estimates, four-fifths of the population of over 5 million still lack electricity in their homes. The total hydroelectric potential of Indonesia is substantial, about 31,000 MWe, with 2,500 MWe in Java, 6,750 MWe in Sumatra, 7,000 MWe in Kalimantan, 5,600 MWe in Sulawesi, and 9,000 MWe in Irian Jaya, the Indonesian part of New Guinea. Altogether, only 650 MWe, or 2.1 percent, has been utilized until now. Unfortunately, most hydroelectric potential is in places that from the point of view of the distribution of the population and the possibility of industrial development are not very promising, a situation comparable to that of China, which also has most of its hydroelectric potential far from population centers.

Nuclear power has already been mentioned. In the mid-1960s, the late President Sukarno startled the world by announcing that Indonesia would soon have nuclear weapons. The claim had no basis in reality. More recently a cabinet-level group has discussed, inconclusively, the future of nuclear energy in Indonesia. Under the sponsorship of the Indonesian National Atomic Energy Agency (BATAN), the French government has looked for uranium in Kalimantan. Neither this nor similar efforts elsewhere appears to have been successful. In preparation for a possible nuclear power plant, a feasibility study was made by the International Atomic Energy Agency and a site was selected in Central Java. It does not seem at present that this will become a very active project in the foreseeable future, despite some pressure from BATAN. Even if Indonesia decides to develop nuclear energy, it is not likely to have its first nuclear power plant on line before 1990.

GOVERNMENT'S ENERGY DEVELOPMENT PLAN

A good overview of Indonesia's energy plans and policies is provided in the official summary of the government of Indonesia's Third Five-Year Development Plan, 1979-1984 (1979 REPELITA III). The section on "Mining, Energy and Power" states:

> Exploration and development of energy resources such as oil and natural gas will receive first priority in Repelita III.

To diversify energy will be accelerated. This exploration and development of energy sources will be implemented in accordance with the national energy policy described below.

To achieve the above objectives, activities which were initiated in the first two Plans will be continued and intensified. These consist of surveys, mapping exploration and exploitation of mineral and energy sources. Research activities to develop a more efficient mining technology, including investigation of mining deposits and alternative processing methods, will be continued. Environmental considerations should be given proper attention during the implementation of above activities. The Plan calls also for institutional and manpower development in the mining sector.

Exploration and development of mineral and energy sources require substantial investment. In this regard the Plan encourages foreign capital to participate in these activities together with domestic capital. Joint ventures in this sector are highly welcome and encouraged.

On the basis of the above objectives and policies, the Plan envisages programmes of production. Oil production is projected to increase from 582 million barrels in 1979/80 to 668 million barrels in 1983/84. In the first two years of this period oil production will be lower than in 1978/79, the last year of Repelita III. This is basically due to a depletion of certain wells and a decline in investment activities in the last few years. In the third year oil production is projected to increase, mainly due to secondary recovery activities using existing wells. These activities are made possible through improved incentives given by the government to oil companies. With a better incentive system and a better investment climate and perhaps more attractive prices in the future, oil exploration is expected to increase.

Production of natural gas during Repelita III is projected to increase from 1,010 billion to 1,595 billion cubic feet. The export of LNG started in Repelita II and this development will continue in Repelita III at a growing rate and thus become an important earner of foreign exchange. Domestic use of natural gas is also increasing, especially in the fertilizer, cement and steel industries.

With development, domestic consumption of oil and natural gas has been increasing and it is expected that this will continue. To meet this demand, processing facilities will be developed and expanded during Repelita III.

In addition, distribution facilities will be expanded and improved.

To utilize more economically the natural gas and by products of oil, four major projects are provided for in the Plan. They are the Olefin Centre in Acch, the Aromatic Centre projects, the Mathanol project and Carbon Black project. To implement these projects, foreign and domestic private capital participation will be encouraged.

Production of coal is projected to increase from 470,000 tons to 1,255,000 tons over the Plan period. Most will be domestically consumed and thus leave oil free for exports. Tin and nickel are projected to increase by respectively 12 percent and 45 percent over the Plan period.

On energy the Plan projects a rapidly growing demand during Repelita III. At the end of the Plan period, energy consumption will be more than double compared with the 1977 figure. In addition the consumption pattern of energy is dominated by oil sources. In 1977 it was estimated that 89.8% of total domestic energy consumption originated from oil, 8.6% from gas, 9.7% from coal and 0.9% from hydro power.

The importance of oil as a major source of public revenue and foreign exchange is stressed in the Plan. Furthermore the Plan calls for action to conserve this non-replaceable source of energy and to change the present pattern of consumption to a more balanced pattern of energy consumption. To implement this, the Plan outlines a national energy policy.

The basic objectives of national energy policy are to assure a gradual shift from a mono-energy economy to a poly-energy one at reasonable prices for the domestic market and to ensure a continuous contribution to the balance of payment and public revenues. To achieve these objectives exploration of conventional energy sources will be accelerated. In remote areas additional incentives will be provided to prospective investors.

The Plan also envisages the stepping-up of research activities to find ways and means to utilize non-conventional energy sources and to develop the technology required. To build up research capabilities the Plan emphasizes manpower development and strengthening of research institutions.

Price and fiscal policy is one of the most important tools for relating and directing energy consumption. Also important is stimulating exploration and development. The

Plan does not specify these policies but it emphasizes their use as part of national energy policy.

Based on the national strategy policy, the Plan undertakes a series of actions consisting of short, medium, and long-term programmes. The short-term programme is to intensify exploration and production of conventional energy sources especially oil and natural gas. The medium-term objective is to diversify energy consumption through efficient use of energy, utilization of coal in power development, utilization of hydro power, increasing the production of firewood and utilization of the waste products of forest industries as a source of energy. Research activities for the utilization of solar energy and nuclear energy will be encouraged. The medium programme also emphasizes the need for capital to meet investment requirements. The long-term programme aims at conservation of energy sources through a more efficient use of energy sources and utilization of replaceable energy sources.

The power sector in Indonesia is still weak in terms of generating capacity. The Plan seeks to increase the power generating capacity by about 3,900 MW, which will approximately double the existing capacity of Repelita II. Transmission and distribution lines will also be expanded. The Plan calls for action to increase the efficiency of power production, distribution and consumption. To encourage rural development, especially rural argo-based industries, the Plan allows for the development of electrification to cover 4,700 villages, one of the effects of which will be to benefit more than 1 million village houses.

The plan stipulates further that power tariffs will be determined by general economic indicators without impeding industrial development, especially the development of small-scale industries.

Investment requirements for power development are quite substantial and become a significant burden on public revenues. In Repelita III, the Plan encourages the participation of private capital in this sector. If this policy succeeds, more resources will be used for rural electrification.

The lengthy preceding statement makes it very clear that Indonesian energy planners have a clear vision of their country's energy future. How much they will achieve will depend not only on political will and administrative skill but also on a number of very important intangibles, of which the most important is the true size of Indonesia's recoverable oil reserves.

It must be pointed out that only 8 of Indonesia's 28 potential oil fields have been explored. Most of those still to be explored are off-shore, where exploration was brought to a dead end by policy decisions taken in the aftermath of the Pertamina crisis. Therefore, it may be indeed excessively pessimistic to work with the figure of 10 billion barrels of recoverable reserves that is usually used in the interna-tional literature.

On the other hand, most knowledgeable foreign experts would be very reluctant to accept the figure of 50 billion barrels, quoted by Indonesian officials, simply because the statistical likelihood of find-ing new giant fields is amall. The only giant fields (those containing at least 500 million barrels of known recoverable oil) found in Indo-nesia are Minas, with 2 billion, and Handil, with 800 million barrels, according to the very detailed survey made by Nehring. He believes that modern petroleum exploration is able to locate efficiently giant fields, and does not include Indonesia among the provinces showing good prospects for the discovery of future giant oil fields.

While the possibility of surprises can never be excluded, as petroleum geology is not an exact science, the 20 Indonesian potential oil fields would each have to contain 2 billion barrels, the size of the Minas field, to make the prediction that Indonesia has 50 billion bar-rels of recoverable oil reserves come true. As stated in the summary of REPELITA III, it is hoped that new discoveries and investment in secondary recovery programs will gradually increase annual produc-tion from 582 million barrels a year (1.57 million b/d) in 1979-80 to 668 million barrels a year (1.83 million b/d) in the last year of the plan, 1983-84.

These targets will not be easily achieved. In the Minas field, operated onshore in Sumatra by Caltex, and in the Handil field, oper-ated offshore of East Kalimantan by Total Indonesie, secondary recov-ery programs are in progress to arrest decline and hold production at current levels. To meet the production goals of REPELITA III is, in the words of the economic section of the U.S. embassy "a formid-able target" as 600 million barrels, the equivalent of one giant field, would have to be found each year in order to maintain existing reserve-to-production ratios.

On the basis of the information reviewed above, it becomes possible to draw some inferences about the role Indonesia is likely to play in the energy future of the United States. The first point to make is that since 1977 the share of Indonesian crude oil exports going to the United States has declined from 37 percent to 29.5 percent, whereas the share to Japan has increased from 48 percent to 57.2 percent. Unlike Japan, the United States has made no special efforts to preserve for Indonesia its share of the American oil import market. For instance, in 1980, Southern California Edison was unable to carry out the contract with Pertamina under which it was purchasing 60,000 b/d of Indonesian crude.

It is suspected that because of its low-sulphur content there were ready takers for that oil in other countries, despite the temporary glut that had developed prior to the Iran/Iraq war. South Korea was known to want to establish a special relationship as a buyer of Indonesian crude, as were two of Indonesia's partners in the Association of Southeast Asian Nations (ASEAN), Thailand and the Philippines. I wonder whether Pertamina can easily shift supplies back to California, once it makes commitments elsewhere. It is quite clear that unlike Saudi Arabia, Kuwait, or Libya, Indonesia has no surge capacity. It is producing all it can within the limits of rational reservoir management.

If there were serious disruptions of the flow of oil from the Middle East, because of military or political circumstances, Indonesia could not step in and fill the gap. If anything, it might have to reduce shipments to Japan and the United States to provide emergency support to its closest political associates, Thailand and the Philippines, which currently import their oil almost exclusively from the Middle East.

Although it is desirable to establish firm long-range relations with Indonesia as a supplier of LNG to California, both for the sake of the California consumer and to help Indonesia to diversify its markets and not become completely dependent on Japan as an exclusive buyer of its LNG, it seems difficult to argue that long-term LNG contracts necessarily would give California the status of preferred customer in case of serious disruption of the flow of Middle Eastern oil. Indonesia is itself importing crude oil from the Middle East (to be processed in its own refineries for domestic consumption) and refined products processed abroad from Middle Eastern crude oil. In case of supply disruptions, Indonesia will obviously allocate a larger proportion of its own production for domestic consumption.

Since major disruptions of Middle Eastern supplies would obviously drive the price of available crude oil to a very high level, Indonesia would probably try to sell as much as possible in the spot market. The likelihood that the United States or even ASEAN countries would be able to purchase on preferential terms does not seem plausible. Indonesia's first responsibility is to its large and poor population and it cannot and should not be expected to act otherwise.

As economic development progresses and domestic demand increases, Indonesia will be forced to reduce oil exports in order to maintain the momentum of its growth process and also to satisfy, for obvious political reasons, the needs of domestic consumers of kerosene, diesel fuel, gasoline, and so forth.

It is sometimes argued that Indonesia will be forced to continue substantial exports of oil to cover imports and the service of its external debt. In its 1980 annual assessment of the Indonesian economy, the World Bank concluded that the combined net value of oil and LNG earnings would "almost certainly" decline in real terms after 1985, although the net value of LNG exports will increase. Its projections indicate that the positive resource balance that amounted to $3.2 billion in 1979 will decline and become a resource gap of $2 billion to $4 billion by 1985, unless Indonesia achieves, inter alia, "realistic oil pricing and substitution policies to further decelerate domestic oil consumption growth and prevent a rapid decline in the exportable oil surplus."

Beyond 1985 the World Bank concluded that projections "become increasingly uncertain." It is assumed that future prospects would depend crucially on projections of oil export revenues. The projections based on a low growth rate of oil production—1.5 percent a year—in conjunction with a moderate growth rate of domestic consumption—8.5 percent a year—would cause the exportable surplus to decline at a rate of about 3 percent a year from 1985.

The World Bank projections assume that oil and LNG net exports, in current prices, will increase from $7.1 billion to $16.8 billion in 1985 and $22.5 billion in 1990. The report does not reveal the basis of the assumption that Indonesia will have $22.5 billion worth of petroleum products available for export in 1990. But taking the REPELITA III production target of 668 million barrels per year (b/y) as basis, a recoverable reserve of 10 billion barrels will last 15 years, or until 1995. How much of that amount will be available for export will depend not only on the energy policy of the Indonesian government but also on the assistance it will receive from abroad to rapidly develop alternative sources of energy. In this respect, the more efficient the assistance that Indonesia will receive from abroad for the implementation of its energy policy, the more petroleum products it will have available for export toward the end of the twentieth century.

NOTES

1. California Energy Commission, Monthly Oil Report, Sacramento, June 1980.
2. Thailand is more fortunate geographically. Underwater pipelines from offshore natural gas from the Gulf of Siam are being developed by Union Oil and others for consumption in the Bangkok area.

V
North American Energy Sources

INTRODUCTION

The three chapters in Part V continue to underscore the complexity of the energy resources picture. Whereas previous chapters focused on non-North American sources where data collection problems, international political uncertainties, and other factors presumably are more pronounced, many of the same problems are encountered in Alaska, Canada, and California. In fact, several additional factors emerge, such as environmental restrictions, federal-state or province relations, conflicting resource requirements such as wildlife, wilderness, and coastal areas, and others. All of these factors and the inherent difficulties associated with projected use patterns and reserves indicate clearly that the simple solutions to U.S. energy problems will not be found in Alaska, Canada, or California. Petroleum is a nonrenewable resource and the articles that follow clearly demonstrate the continuing impact of that simple, but too often neglected, fact.

11

THE OUTLOOK FOR ALASKAN NORTH SLOPE CRUDE OIL PRODUCTION, 1980–2000

Arlon Tussing

This chapter considers the outlook for future crude oil production in Arctic Alaska as one element enhancing California's and the United States' energy security and also, ironically, as a source of energy insecurity. The projections of North Slope oil production were prepared originally as part of an analysis concerning the demand for a new West-to-East crude oil pipeline system. For this reason, the chapter also reflects briefly on how the Alaskan production outlook affects the viability of some of the competing pipeline proposals.

Table 11.1 shows the best estimates of future crude oil production in Arctic Alaska between 1980 and 2000, based on information available in early 1980. Production will most likely crest at about 1.9 million barrels per day (b/d) in the mid–1980s and fall to about 800,000 b/d by the year 2000. There is, however, about a 5 percent probability that peak production will never exceed the current level (less than 1.6 million b/d) and will decline to as low as 400,000 b/d by the end of the century, and about a 5 percent probability that production will exceed 2 million b/d by 1986 and reach a peak of more than 4 million b/d before the year 2000. Table 11.1 presents the forecasts for each year at 95, 50, and 5 percent confidence levels. The "most likely" projections tend to be substantially lower than recent forecasts from the other sources shown in Table 11.2 later in the chapter.

TABLE 11.1

Alaska North Slope Crude Oil Production, 1980-2000
(thousand barrels per day)

	Confidence Level		
	Low (95%)	Most Likely (50%)	High (5%)
1980	1,484	1,500	1,560
1981	1,484	1,500	1,560
1982	1,484	1,539	1,597
1983	1,452	1,558	1,643
1984	1,452	1,690	1,815
1985	1,500	1,771	1,950
1986	1,530	1,808	2,133
1987	1,585	1,906	2,332
1988	1,410	1,745	2,440
1989	1,092	1,465	2,798
1990	910	1,377	2,898
1991	759	1,295	3,270
1992	750	1,289	3,566
1993	720	1,279	3,541
1994	648	1,185	3,451
1995	584	1,112	3,429
1996	569	1,094	3,820
1997	521	1,013	4,055
1998	476	935	4,220
1999	433	867	4,040
2000	394	791	3,844

Source: Compiled by author.

NORTH SLOPE CRUDE OIL AND WEST COAST ENERGY SECURITY

In 1978, a surplus of crude oil appeared on the West Coast of the United States as a result of new production in Arctic Alaska. The existence of this surplus depressed wellhead prices of crude oil in both Alaska and California below their world market price equivalents, thereby reducing Alaska's tax and royalty income and causing the shutting-in of established heavy oil production in California. On the initiative of the Alaska legislature, the two states created a joint interagency study group to deal with these problems.

The study group considered several measures to mitigate the surplus, including an effort to get licenses to export North Slope crude oil and California residual oil to the Far East, and support by the two states to one or more of the proposed West-to-East crude oil piplines. The group devoted its greatest attention, however, to possible state or federal incentives for California refiners to use more Alaska and California crude oil. Using more local crude oil on the West Coast, it was believed, would not only benefit the two states economically, but, by displacing Eastern Hemisphere oil imports, it would make West Coast oil supplies less vulnerable to future overseas crises.

The world oil price upheaval that followed the Iranian revolution, the Department of Energy's (DOE) special treatment of Alaska oil under the price control program, and the legal prohibition on exports to the Far East all combined to create an unexpectedly powerful incentive for West Coast refiners to run Alaska oil. California refineries absorbed considerably more North Slope crude oil in 1979 and early 1980 than anybody in the industry or in government—including the members of the study group—had publicly anticipated.

During the first half of 1980, 43 percent of California's total petroleum supply came from Alaska, and 40 percent came from a single reservoir on the North Slope, through a single pipeline and a single marine terminal at Valdez. As a result, the most vulnerable point in the West Coast's energy supply today may, ironically, be the very reliance on North Slope crude oil that we were seeking to strengthen only two years ago.

A break in the Trans-Alaskan pipeline (TAPS) caused by an earthquake, volcanic eruption, accident, or sabotage could be repaired in a few days. But restoration of a disabled or destroyed pump station on the pipeline would require weeks, while the Yukon River crossing or the Valdez terminal could conceivably be put out of service for a few months to a year or more.

While these events may not be as likely as another international or civil conflict in the Middle East, they are certainly plausible. The regional consequences of interrupting all shipments through TAPS would, moreover, be deeper and felt much more quickly than any likely development in, for example, Indonesia or Saudi Arabia. California would begin drawing down stocks within seven days after an interruption of the supply from Alaska because tankers take about that long to travel from Valdez to California ports. By contrast, interruptions of supplies from the Middle East or Indonesia would not be felt for several weeks. The thought of a prolonged TAPS shutdown

coming at the same time as a new Middle Eastern crisis is truly hor-
rifying, and has inspired at least one disaster novel.

The seriousness of the problem and its potential solutions both
depend, at least in part, upon the outlook for future crude oil production from Alaska's North Slope. Specifically, will the West Coast's
dependence on TAPS tend to increase as additional production on the
Arctic Slope further displaces imports and (perhaps) the less-desirable
grades of California crude oil? Or is the current preponderance of
North Slope crude oil in the region's supply only a temporary "bubble"
that will shrink as Prudhoe Bay production declines and other supplies—Alaska or California offshore, California heavy crudes, or increased imports—take its place?

NORTH SLOPE FORECASTS AND TRANSCONTINENTAL
PIPELINE PROPOSALS

The same questions are at the heart of the recent controversies
over proposed East-to-West crude oil pipelines and their associated
port facilities. Whether a new transportation system is economically
warranted at all, and if so which one, also depends powerfully on the
volume of crude oil to be shipped from Arctic Alaska during the rest
of this century.

The Northern Tier pipeline company, which would build a line
from Puget Sound to Minnesota, and the Trans-Mountain pipeline,
which would connect Puget Sound to Edmonton, justify their respective
proposals mainly as an outlet for surplus Alaska oil. Sohio owns most
of the surplus and suffers the worst transport burden of all the North
Slope producers. In 1978, Sohio abandoned its own Long Beach-to-
Texas pipeline project on the ground that there would not be enough
surplus oil left at Prudhoe Bay when the pipeline was completed to
warrant its capital expense. Likewise, the Foothills group, which
proposes to build a totally overland line linking TAPS at Big Delta to
the existing continental pipeline grid in Alberta, has temporarily with-
drawn its applications, stating that currently proved Alaska reserves
would not support any of the proposed pipeline systems. Foothills
maintains that its proposal would provide the most efficient transportation for Alaska oil surplus to West Coast demand only when and if
new discoveries assure that TAPS throughput will equal or exceed
present levels for 10 to 20 years.

In reviewing the materials published by government agencies,
pipeline sponsors, and other industry groups, one is struck by the
differences among their forecasts of future North Slope production,
by the fact that none of the sources show the full assumptions behind
these forecasts, and above all, by many of the forecasters' seeming

unawareness or disregard of the specific uncertainties involved in making such predictions.

Table 11.2 summarizes some of the most widely cited production forecasts published since 1975 for 1980, 1985, 1990, and 2000, from industry and governmental sources. For each source, the table shows the cumulative amount of crude oil that would have been produced at the end of each forecast year, and the total volume of reserves that would have to be discovered and developed for production by that time, on the standard assumption that remaining reserves must equal at least 10 times the final year's production.

Consider, for example, the 1979 forecast that Pace, Inc., prepared for Butler Associates, Inc., as part of the supporting documents for the Northern Tier pipeline application and that projects Alaska oil production rising from 1.2 million b/d in 1978, to 1.7 million b/d in 1985, 2 million b/d in 1990, and 3.3 million b/d in the year 2000. On first impression, these forecasts are certainly plausible, but they are hard to reconcile with some of the assumptions on which they are based.

Pace estimates that Alaska had 30 billion barrels of recoverable oil left to be discovered in 1979, both onshore and offshore. Together with about 7 billion remaining barrels of proved reserves at Prudhoe Bay and in Cook Inlet, the state's total crude oil resource base would be about 37 billion barrels. Pace's production forecasts for 1980 through 2000 cumulate to about 18 billion barrels. If it is assumed, conservatively, that proved reserves at the end of the year 2000 would have to equal at least ten years' production at the rate achieved in that year, another 12 billion barrels would have to be discovered and developed by the end of the century in addition to the 18 billion barrels previously consumed, for a total of at least 30 billion barrels.

Out of the state's 37 billion barrels of recoverable oil-in-place as of 1979, in other words, Pace assumes that 30 billion barrels— 81 percent—will have been discovered and developed for production in the next 20 years. But there is not the slightest chance that all of the promising onshore and offshore areas of Alaska will even be leased, let alone explored and developed in this century.

OVERVIEW OF ALASKA PETROLEUM DEVELOPMENT

Most other recent forecasts, like the Pace report, show substantial increases in oil production from Arctic Alaska over the next 20 years, but in none of them is it clear just how and where those levels of production can be achieved. The Appendix at the end of this chapter contains admittedly rough area-by-area production forecasts, and their general outlook is summarized below.

TABLE 11.2

Alaska Oil Production Estimates

Source of Forecast	Annual Production million barrels/day				Cumulative Production[a] billion barrels to			Reserves Required[b] billion barrels to		
	1980	1985	1990	2000	1985	1990	2000	1985	1990	2000
DeGolyer & McNaughton, 1975	1.5	1.9	2.0	—[c]	4.5	8.1	—	11.4	15.4	—
Rand Corporation, 1975	1.6	2.2	2.8	—	4.9	9.6	—	12.9	19.8	—
University of Texas, 1976	1.6	2.0	2.0	—	4.8	8.5	—	12.1	15.8	—
Sohio, 1976	1.5	2.0	—	—	4.6	—	—	11.9	—	—
Federal Energy Administration, 1976										
"pessimistic"	1.2	1.8	1.8	.8	4.1	7.4	12.0	10.7	14.0	14.9
"business as usual"	1.6	2.8	2.4	.7	5.6	10.3	14.6	15.8	19.1	16.6
"optimistic" (w/o NPRA)*	2.3	3.3	2.2	.2	7.0	11.9	14.5	19.0	20.0	15.3
"optimistic" (with NPRA)	2.3	4.5	2.8	.3	8.1	14.3	18.2	24.5	24.5	19.3
Department of Energy, E.I.A.,** 1978	1.2	1.6	1.6	—	3.9	6.8	—	9.8	12.9	—
A. D. Little, Inc.										
Petroleum Transport Syst., 1977	1.6	1.8	—	—	5.5	—	—	12.1	—	—
California Clean Fuels Study, 1978										
most likely[d]	1.7	2.8	2.9	—	5.7	11.6	—	15.9	22.2	—
high[d]	2.4	3.1	3.6	—	6.8	13.0	—	18.0	26.1	—
Pace, Inc.										
Low Sulfur Crude Oil Study, 1978	1.6	2.0	2.5	2.4	4.8	9.0	18.0	12.1	18.1	36.1
Butler Bros. for N. Tier, 1978	1.6	1.9	2.3	—	4.7	8.6	—	11.8	16.0	—
Butler Bros. for N. Tier, 1979	—	1.6	1.8	1.3-1.6	4.3	7.4	12.3-13.6	10.1	13.7	17.0-19.4

California Energy Commission, 1979	—	1.6	1.8	2.5	4.3	7.4	15.3	10.1	13.7	24.4
Department of Energy										
Office of O & G Policy, 1979										
low	1.2	1.5	1.5	1.8	3.8	6.6	12.4	9.2	12.0	24.6
high	1.2	1.8	2.2	2.2	4.1	7.9	15.9	10.7	15.9	31.8
Energy Inform. Adm., 1979										
low	—	1.5	1.4	1.5	3.9	6.6	11.9	9.6	11.9	17.2
high	—	1.5	1.5	1.8	3.9	6.7	12.7	9.6	12.4	19.4
Arlon R. Tussing & Assoc., Inc., 1980										
low (95%)	1.5	1.1	.9	.4	4.0	6.4	8.5	9.0	9.7	10.0
most likely (50%)	1.5	1.8	1.4	.8	4.3	7.3	11.2	10.8	12.3	14.1
high (5%)	1.6	1.8	3.0	3.8	4.5	10.2	23.8	11.6	20.7	34.0

a Annual data interpolated if not available from source, and include 0.8 billion barrels produced in 1977–79.

b Cumulative production plus remaining reserves equal to 10 times final year's production.

c Data not available.

d Alaska onshore and new offshore.

Note: Alaska North Slope or "North Alaska" unless otherwise indicated.

Source: Compiled by the author.

*Naval Petroleum Reserve Act (N.P.R.A.).

**United States Energy Information Agency (E.I.A.).

Reserves in two areas, Prudhoe Bay and Cook Inlet, are developed and are now producing. The production from developed fields in Cook Inlet has been declining at a rate of about 17 percent a year. New discoveries may conceivably stem that decline, but barring unforeseen and exceptional good luck or new discoveries, the established producing areas in Cook Inlet are already well past their peak. Production from the main reservoir at Prudhoe Bay (the Sadlerochit formation) will continue at about 1.5 million b/d for another four or five years, and then begin to decline at roughly 12 percent a year. These estimates are uncertain because some disagreements remain among the various producers and the state's engineers and consultants about the reservoir's long-term behavior, and there may yet be a serious conflict over when to begin artificial waterflood (which benefits the state, but for which the producers must pay).

Two other reservoirs have been found in the Prudhoe Bay field, and either may prove to be a "super-giant," that is, to contain more than 1 billion barrels of recoverable oil. But little information about them is currently in the public domain, and it is unlikely that they are large enough or will be developed rapidly enough to offset the production decline in the main reservoir. There have been at least two, and perhaps three, other major discoveries on the North Slope in the last two years. One of them, at Point Thomson, could also contain over 1 billion barrels, but again development is a long way off. The only development scheduled for the period before 1985 is at the Kuparuk reservoir in the Prudhoe Bay field. ARCO anticipates production from this formation of something in the vicinity of 100,000 b/d by 1984.

The discovery of a supergiant field is a rare and random event. The Prudhoe Bay field is already the largest ever discovered in the United States or Canada. The number of barrels that have to be discovered between now and the year 2000 to validate the highest production projections imply an annual rate of discovery in Alaska alone greater than has ever been achieved in the rest of North America over a similar period of time. The odds do not favor any development combination in known reservoirs or on already leased prospects that could more than offset the certain decline of production in the main Prudhoe Bay reservoir between now and the end of the century.

There are no plans now for ever leasing all of Alaska's prospective oil and gas acreage. Some of the best exploration targets are in places like the William O. Douglas Arctic National Wildlife Range, which the president and the House of Representatives want to designate as a permanent wilderness. Even the Senate bill, which is backed by the state of Alaska and the prodevelopment forces, would require a five-year federal study followed by further congressional action before leasing could proceed.

Other important exploration targets are in sensitive wildlife areas which, while they are not officially designated wilderness areas, federal and state land managers are unlikely to lease in the foreseeable future. And every proposed lease sale is certain to be contested by someone—fishermen, native groups, environmentalist organizations, or local governments. As a result, some sales will undoubtedly be cancelled and others will be delayed just as the 1979 federal and state lease sales in the Beaufort Sea are both tied up in the courts.

STATE ATTITUDES TOWARD DEVELOPMENT

Alaska's political rhetoric seems, particularly from the outside, to be militantly prodevelopment. The rhetoric can be misleading, however, as the state's effective political majority is probably as concerned about environmental protection as economic development, and tends to be rather skeptical about policies designed to accelerate industrialization. There is little enthusiasm, at any rate, for simply realizing more cash bonuses from additional oil and gas leasing—the state's program to accelerate leasing on its own lands has no perceptible public constituency (except in the oil industry itself, which is still generally regarded as an outside influence).

In three, five, or ten years, perhaps, when the state's ability to spend money has caught up with its Prudhoe Bay revenues, or when those revenues begin to turn down, it may still appear easier and more immediately effective for the state to squeeze a little more money out of the oil companies on the properties that they have already developed than to accept the delays and the risks of failure in a fiscal strategy that depends on new discoveries.

LOGISTICAL CONSTRAINTS

Even if all of the prospective Alaska onshore and offshore acreage were now available for development, moreover, there is no physical way in which the oil industry could explore most of that acreage before the turn of the century. Finally, in much of Alaska, climate and transportation problems make it unlikely that any commercial production could take place for ten years after a lease sale.

TREATMENT OF UNCERTAINTY

Major uncertainties that have to be considered in forecasting crude oil production from Arctic Alaska include the future strategy of the operating companies and regulations of the state with regard to management of the Prudhoe Bay reservoir, and the actual behavior of that reservoir over its productive life; development strategy and performance with regard to three large but yet undeveloped reservoirs or fields in the vicinity of Prudhoe Bay; future state and federal policy with regard to Arctic oil and gas leasing, both onshore and offshore; congressional action directing, permitting, postponing, or forbidding oil and gas leasing on the National Petroleum Reserve in Alaska (NPRA) and the William O. Douglas Arctic National Wildlife Range (ANWR); the timing, number, and size of new discoveries (if any) in these areas, and the timing of development and production from such discoveries.

Many of these variables are not subject to rigorous geological or economic analysis, so that predicting future production from a particular unexplored area is essentially a guess. Some of the uncertainties, moreover, are of kinds that do not allow the quality of such guesses to be improved greatly by the use of elaborate forecasting methodologies. Because of these uncertainties, production from each prospect or area (except for the main Prudhoe Bay reservoir) could plausibly range from zero to very substantial figures (more than 500,000 b/d in NPRA, ANWR, or the Beaufort Sea). Thus, one's judgment about the relative probability of different levels of production is almost as important as one's guess regarding the "most likely" level.

This chapter has attempted to remedy the main shortcomings of existing forecasts by explicitly stating both the assumptions and the subjective probabilities attached to various levels of production. The forecasts are preliminary and expected to be revised from time to time.

The estimates in Table 11.1 do not rest on any proprietary information or original research; rather, they reflect information in the public domain and judgments regarding future leasing schedules, the relative resource endowment of various areas, producer intentions, and development economics. The following Appendix shows separate production forecasts for the Prudhoe Bay reservoir, the Kuparuk and Lisburne formations of the Prudhoe Bay field, the Point Thomson field, the area of the 1979 state-federal lease sale in the Beaufort Sea, NPRA, and ANWR. A forecast for "all other areas" covers possible new discoveries on other leased or unleased state onshore lands (except for an anticipated drainage sale on the Point Thomson structure, which is included above), state or federal leases offshore of NPRA or ANWR, and land owned by the Arctic Slope Regional Corporation (ASRC). It does not include onshore or offshore state, federal, or ASRC lands west of NPRA.

APPENDIX: METHODOLOGY AND
INDIVIDUAL AREA FORECASTS

The forecasts in Table 11.1 are of total crude oil production
from all Alaska North Slope areas included in the Appendix tables, at
95 percent, 50 percent, and 5 percent confidence levels. Thus, Table
11.1 indicates that it is believed there is a 95 percent probability that
production will be at least 1.124 million b/d in 1985; that the probabilities are about equal—50 percent—that actual production will be less
than, or greater than, 1.771 million b/d; and that there is a 5 percent
probability that total production will equal or exceed 1.950 million b/d.

It is possible that production in a given year will be much less
or much greater than the minimum and maximum figures in the table,
perhaps because of an irreparable pipeline failure or discovery of one
or more new Prudhoe Bay-size oil fields. The table shows, however,
that the likelihood of either event is remote—specifically as less than
5 percent.

The production forecasts for the various areas shown in the
Appendix tables have been aggregated by the "Monte Carlo" technique
and assume that the expected production for each area in each year is
independent of all other areas and all other years, and that the individual probability distributions are lognormal, except in the case of
the Prudhoe Bay Sadlerochit reservoir where the distribution is made
up of two lognormal segments joined at the 50 percent confidence
point. Production rates in each distribution are truncated to zero at
the low end, at the point where the confidence level equals the "probability of commercial production" shown in the tables.

The total production estimated for each confidence level is not
the sum of the production estimates shown for the various areas at
that level. (For example, in each area except the Prudhoe Bay field,
there is at least a 5 percent likelihood that there will be no production
at all before the year 2000. But it does not follow that there is a 5 percent probability that there will be no production from any of them.)
Thus, the 95 percent probability level estimate for total production
is significantly greater than the sum of the 95 percent estimates for
the individual areas, and the 5 percent probability estimate for total
production is significantly smaller than the sum of the 5 percent estimates for the individual areas. For mathematical reasons too complex
to be set out here, the 50 percent confidence estimate of total production can also be expected to differ from the sum of the individual 50
percent estimates.

Prudhoe Bay Unit

All 1980 production in Arctic Alaska was in the Jurassic-Tri-
assic sandstones of the Sadlerochit reservoir at Prudhoe Bay. The

TABLE A11.1

Prudhoe Bay Sadlerochit Reservoir, Crude Oil Production,
1980-2000
(thousand barrels per day)

	Probability of Commercial Production (percent)	Confidence Level		
		95%	50%	5%
1980	98%	1,483	1,500	1,560
1981	98	1,483	1,500	1,560
1982	98	1,483	1,500	1,560
1983	98	1,448	1,480	1,600
1984	98	1,448	1,480	1,600
1985	98	1,084	1,490	1,600
1986	98	921	1,540	1,600
1987	98	783	1,540	1,600
1988	98	666	1,370	1,500
1989	98	566	1,050	1,500
1990	98	481	800	1,365
1991	98	409	765	1,242
1992	98	347	570	1,130
1993	98	295	450	1,029
1994	98	251	370	936
1995	98	213	310	852
1996	98	181	250	775
1997	98	154	200	705
1998	98	131	160	642
1999	98	111	130	584
2000	98	95	100	532

Source: Compiled by author.

state estimated that total remaining reserves in the unit (including
the Sadlerochit and much smaller Sag River reservoirs) at the end of
1980 would be 7 billion to 8.2 billion barrels, with a most likely value
of 7.8 billion barrels.

The producing companies plan to maintain the 1.5 mb/d

output attained at the end of 1979 for as long as practical. The "most likely" (50 percent confidence) estimates in Appendix Table 11.1 are from the Doschers Group, Inc., "Analysis of the Results of the Numerical Simulation of the Sadlerochit Reservoir Prudhoe Bay Field," run 233 in which waterflood begins in the eighth year and 2 billion cubic feet a day (BCF/d) gas production begins in the ninth year according to a February 3, 1980, report by the Alaska State Legislature, Joint Gas Pipeline Committee. Substantial uncertainty remains, however, about how much oil will ultimately be recovered, and with it the number of years the reservoir can (or will) be developed to produce at peak capacity and the rate of decline thereafter. Statements by the producers imply that the peak production of 1.5 million b/d will be maintained for only 5.6 years, but this is a conservative projection if, as likely, the state compels (or allows adequate incentives for) the companies to institute early water flooding and other measures to enhance recovery.

Summary: Prudhoe Bay Sadlerochit Reservoir

	Confidence Level		
	95%	50%	5%
Peak production	1484 mb/d	1540 mb/d	1600 mb/d
Peak reached in year	1979	1979	1983
Period of peak production	6 years	8 years	10 years
Subsequent decline rate	16%	12%	9%

Kuparuk River Formation

ARCO has announced its intention to develop the Kuparuk River formation in Cretaceous rocks to the west of and above the Sadlerochit pool in the Prudhoe Bay field to produce 60 million b/d in 1982 and 100 million b/d in 1984. The 50 percent probability column reflects this intention. The state estimates the Kuparuk reservoir's reserves to be between 400 million and 1.3 billion barrels, with a most likely figure of 750 million barrels. There is a reasonable chance that the formation might produce as much as 200 million b/d at its peak. On the other hand, there is a significant likelihood that high costs or other problems will prevent this reservoir from producing any oil in commercial volume before the year 2000.

TABLE A11.2

Kuparuk River Formation: Crude Oil Production,
1980–2000
(thousand barrels per day)

	Probability of Commercial Production (percent)	Confidence Level		
		95%	50%	5%
1980	0	0	0	0
1981	0	0	0	0
1982	60	0	60	60
1983	60	0	80	80
1984	70	0	100	120
1985	70	0	100	150
1986	80	0	100	200
1987	90	0	100	200
1988	90	0	100	200
1989	90	0	100	200
1990	90	0	90	200
1991	90	0	81	200
1992	90	0	73	200
1993	90	0	66	200
1994	90	0	59	200
1995	90	0	53	200
1996	90	0	48	184
1997	90	0	43	169
1998	90	0	39	156
1999	90	0	35	144
2000	90	0	31	132

Source: Compiled by author.

Summary: Kuparuk River Formation

	Confidence Level		
	95%	50%	5%
Peak production	none	100 mb/d	200 mb/d
Peak reached in year	none	1984	1986
Period of peak production	none	6 years	10 years
Subsequent decline rate	none	10%	8%

Lisburne Formation, Point Thomson Field,
and Related Areas

The Lisburne limestones below and to the east of the Prudhoe
Bay field, and the area of Exxon's two discovery wells near Point
Thomson (about 90 km to the northeast of Prudhoe Bay) and Flaxman
Island have been combined here with the nearby and possibly related
plays in the Saganavirtok Delta and in the Duck Island area. Indications
thus far are that these prospects include at least two and perhaps
more "giant" reservoirs, with recoverable reserves in the hundreds
of millions of barrels. The state estimates total reserves in the
Prudhoe Bay Lisburne Reservoir, the Sag Delta area, and the Duck
Island area at 460 million to 975 million barrels, with 600 million
barrels the most likely; for the Point Thomson-Flaxman Island reser-
voir(s), the state estimates reserves of 400 million to 900 million
barrels, with a most likely value of 600 million.

Some geologists are more optimistic about one or both of these
prospects, believing that they have a reasonable chance to be "super-
giants" (on the billion-barrel order of magnitude), yet no industry
reserve estimates or development plans have been published for either.
Geophysical and geological information regarding both prospects has
great competitive value for companies that won leases in the December 1979 federal-state lease sale (or that intend to bid in future sales
of adjacent tracts). Development plans or reserve estimates are not
likely to be published until after completion of these sales and the
subsequent trading of lease rights. The two areas have been combined
for forecasting purposes and it is assumed that each will most likely
be capable of producing about 100 million b/d by 1985. There is,
however, a substantial possibility that economics or other considerations will prevent the development of either field before the end of
the century and a comparable likelihood that the two could between
them produce 400 million b/d or more.

TABLE A11.3

Lisburne Formation, Point Thomson Field, and
Related Areas: Crude Oil Production, 1980-2000
(thousand barrels per day)

	Probability of Commercial Production (percent)	Confidence Level		
		95%	50%	5%
1980	0	0	0	0
1981	0	0	0	0
1982	0	0	0	0
1983	60	0	50	50
1984	70	0	100	125
1985	70	0	200	225
1986	70	0	200	300
1987	80	0	200	400
1988	80	0	200	400
1989	80	0	200	400
1990	90	0	200	400
1991	90	0	180	400
1992	90	0	162	400
1993	90	0	146	400
1994	90	0	131	400
1995	90	0	118	400
1996	90	0	106	400
1997	90	0	96	368
1998	90	0	86	339
1999	90	0	77	311
2000	90	0	70	286

Source: Compiled by author.

Summary: Lisburne Formation, Point Thomson Field,
and Related Areas

	Confidence Level		
	95%	50%	5%
Peak production	none	200 mb/d	400 mb/d
Peak reached in year	none	1985	1987
Period of peak production	none	6 years	10 years
Subsequent decline rate	none	10%	8%

1979 Beaufort Sea Lease Sale Area

A joint federal-state lease sale was held in December 1979, offshore of the Prudhoe Bay area. The 50 percent probability figures for production in this sale area were taken unadjusted from the "Beaufort Sea Intermediate Case," in the draft environmental impact statement (DEIS) prepared for the lease sale. The DEIS "minimum," "intermediate," and "maximum" cases all reflected discovery and development of one oil field, of different sizes. It is believed that there is a substantial probability that the leases will be invalidated, or that the area will not produce commercially economic oil discoveries, or if such discoveries take place, that engineering, environmental, or regulatory problems or litigation will forestall production until after the year 2000. The 5 percent confidence forecasts assume there would be three commercial discoveries in the sale area, one of "intermediate" size commencing production in 1991, one of "maximum" size two years later, and a second "intermediate" size field after another two years.

Summary: 1979 Beaufort Sea Lease Area

	Confidence Level		
	95%	50%	5%
Peak production	none	151 mb/d	548 mb/d
Peak reached in year	none	1993	1996
Period of peak production	none	6 years	5 years
Subsequent decline rate	none	n/a*	n/a

*Data not available.

TABLE A11.4

1979 Beaufort Sea Lease Area: Crude Oil Production,
1980–2000
(thousand barrels per day)

	Probability of Commercial Production (percent)	Confidence Level		
		95%	50%	5%
1980	0	0	0	0
1981	0	0	0	0
1982	0	0	0	0
1983	0	0	0	0
1984	0	0	0	0
1985	0	0	0	0
1986	0	0	0	0
1987	0	0	0	0
1988	0	0	0	0
1989	0	0	0	0
1990	0	0	0	0
1991	60	0	4	4
1992	70	0	113	113
1993	80	0	148	271
1994	90	0	151	356
1995	90	0	151	402
1996	90	0	151	513
1997	90	0	148	545
1998	90	0	145	548
1999	90	0	142	548
2000	90	0	132	532

Source: Compiled by author.

National Petroleum Reserve in Alaska

Projections of production for the National Petroleum Reserve
in Alaska (NPRA) are derived from the Interior Department's draft
report of the 105(b) Economic Policy Analysis (July 31, 1979). Fore-

TABLE A11.5

National Petroleum Reserve in Alaska: Crude Oil
Production, 1980-2000
(thousand barrels per day)

	Probability of Commercial Production (percent)	Confidence Level		
		95%	50%	5%
1980	0	0	0	0
1981	0	0	0	0
1982	0	0	0	0
1983	0	0	0	0
1984	0	0	0	0
1985	0	0	0	0
1986	30	0	0	91
1987	40	0	0	183
1988	50	0	66	366
1989	60	0	132	731
1990	70	0	132	731
1991	80	0	132	731
1992	80	0	132	731
1993	80	0	132	731
1994	80	0	132	731
1995	80	0	132	731
1996	90	0	119	731
1997	90	0	107	731
1998	90	0	96	731
1999	90	0	87	673
2000	90	0	78	619

Source: Compiled by author.

casts are from policy scenario #2, which assumes traditional leasing
and development procedures with lease sales beginning in 1983. The
summary of outputs from this scenario shows an "average value,"
and a "5 percent value" of "sample daily flow rates" at 132 million
and 731 million b/d respectively. These figures have been taken as

234 / THE UNITED STATES AND WORLD ENERGY SOURCES

peak production rates at the 50 and 5 percent probability levels, and
have assumed that there is a substantial chance that there will be no
commercial production from NPRA in this century.

Summary: National Petroleum Reserve in Alaska

	Confidence Level		
	95%	50%	5%
Peak production	none	132 mb/d	731 mb/d
Peak reached in year	none	1989	1989
Period of peak production	none	7 years	10 years
Subsequent decline rate	none	10%	8%

William O. Douglas Arctic National Wildlife Range

The northern part of the William O. Douglas Arctic National
Wildlife Range (ANWR) is regarded by some government and industry
geologists as the more promising petroleum prospect under U.S.
jurisdiction. A U.S. Geological Survey report has characterized it
as "highly prospective for accumulations up to 5×10^9 barrels."

The wilderness status of ANWR is one of the highest priority
objectives of the national environmental protection organizations,
however, and the range has been proposed by the president and secretary of the interior for permanent wilderness status that would preclude oil and gas exploration. The U.S. House of Representatives has
adopted legislation to this effect, while the companion bill in the
Senate would require a five-year geological and geophysical study by
the government, followed by congressional action, before any leasing
would be permitted in ANWR. It is believed, therefore, that there is
a reasonable, but less than even, chance that some crude oil production will take place there before the end of the century; that petroleum
exploration would result in discovery of at least one supergiant field,
and that there is a 5 percent likelihood that production from ANWR
could reach 500 million b/d by 1992.

William O. Douglas Arctic National Wildlife Range:
Crude Oil Production, 1980–2000
(thousand barrels per day)

	Probability of Commercial Production (percent)	Confidence Level		
		95%	50%	5%
1980	0	0	0	0
1981	0	0	0	0
1982	0	0	0	0
1983	0	0	0	0
1984	0	0	0	0
1985	0	0	0	0
1986	0	0	0	0
1987	0	0	0	0
1988	0	0	0	0
1989	0	0	0	0
1990	0	0	0	0
1991	30	0	0	125
1992	30	0	0	500
1993	40	0	0	500
1994	40	0	0	500
1995	40	0	0	500
1996	40	0	0	500
1997	40	0	0	500
1998	40	0	0	500
1999	40	0	0	500
2000	40	0	0	500

Source: Compiled by author.

Summary: William O. Douglas Arctic National Wildlife Range

| | Confidence Level | | |
	95%	50%	5%
Peak production	none	none	500 mb/d
Peak reached in year	none	none	1992
Period of peak production	none	none	9 years
Subsequent decline rate	none	none	n/a*

All Other Areas

"All other areas" in Appendix Table 11.7 include both leased and unleased state land in the vicinity of Prudhoe Bay; state and federal lands offshore of NPRA and ANWR, plus onshore and offshore lands between them not scheduled for lease in 1979; and land owned by the Arctic Slope Regional Corporation (ASRC). Chevron, Hamilton Brothers, and Conoco have each announced one discovery well on state land in the Milne Point area, where the state estimates reserves of 30 million to 80 million barrels, with a most likely value of 45 million. Mobil has one discovery well and Hamilton Brothers two in the Gwyrdyr Bay area, where the state estimates reserves of 50 million to 100 million barrels, with a most likely value of 80 million. Further discoveries are likely in both areas, but in their absence, these prospects may be too small to develop profitably in the Arctic.

No other usable reserve estimates or forecasts exist for individual North Slope areas except those previously listed, and although the region contains many attractive exploration targets, there is a substantial possibility that none of them will actually produce crude oil before the year 2000. More likely, there will be several small discoveries; assuming a 50 percent probability, production will reach 100 million b/d by 1991 and remain at that level for at least a decade. There is some potential, particularly offshore of ANWR and NPRA, for discovery of supergiant oil accumulations, and thus it is assumed that there is a 5 percent probability that total production from these areas could reach 500 million b/d by 1994.

---

*Data not available.

TABLE A11.7

All Other Areas: Crude Oil Production, 1980–2000
(thousand barrels per day)

	Probability of Commercial Production (percent)	Confidence Level		
		95%	50%	5%
1980	0	0	0	0
1981	0	0	0	0
1982	0	0	0	0
1983	0	0	0	0
1984	0	0	0	0
1985	40	0	0	50
1986	50	0	50	75
1987	60	0	50	75
1988	60	0	50	100
1989	70	0	50	100
1990	70	0	50	100
1991	80	0	100	300
1992	80	0	100	300
1993	90	0	100	300
1994	90	0	100	500
1995	90	0	100	500
1996	95	0	100	500
1997	95	0	100	500
1998	95	0	100	500
1999	95	0	100	500
2000	95	0	100	500

Source: Compiled by author.

Summary: All Other Areas

	Confidence Level		
	95%	50%	5%
Peak production	none	100 mb/d	500 mb/d
Peak reached in year	none	1991	1995
Period of peak production	none	10 years	5 years
Subsequent decline rate	none	n/a*	n/a

*Data not available.

12

CANADIAN OIL SUPPLIES

R. D. Hall

Following the OPEC price increases in 1973, the control and pricing of oil in Canada became a major issue between the federal and provincial governments. To protect its ownership rights to the oil, the Alberta government passed the Petroleum Marketing Act, which established the Alberta Petroleum Marketing Commission. It sells all royalty oil on behalf of the province—about one-third of the total conventional oil production within the province. The commission also has the exclusive agent to sell the lessee's share of the oil produced from crown lands. That represents about one-half of total production. The balance of the production is freehold oil and is sold by the owner or lessee. The pricing of that oil almost exactly follows the pricing of the crown oil.

During 1975, after the act was established, the province signed an accord with the federal government regarding natural gas pricing that now (1980) fixed the price in Toronto at $2.30 per million British thermal units (Btu). The commission also administers the Natural Gas Pricing Agreement Act, which establishes a field price in Alberta, but the field price is dependent on the federal/provincial accord for the natural gas pricing.

An important aspect of this administration is the distribution of the revenue from exports. The difference between the fixed Canadian natural gas price and that paid for natural gas exported to the United States is distributed to all gas producers, rather than just those who happen to be fortunate enough to be selling to an exporting company. Most natural gas exports come from Alberta, though British Columbia contributes a fair share. The flowback from this averaged about $.47 per million Btu in 1979. The latest increases in the Canadian border

TABLE 12.1

Consumption and Production of Synthetic and Conventional
Crude Oil and Pentanes Plus, 1979
(thousand b/d)

Province	Consumption	Production
Alberta	274	1,399
Atlantic Provinces	326	0
British Columbia	166	41
Manitoba	29	10
Northwest Territories	3	3
Ontario	600	2
Quebec	509	0
Saskatchewan	42	160

Source: Compiled by the author.

price to US $4.47 per million Btu has resulted in the doubling of the
price adjustment to $1.01.

Canada currently produces about 1.85 million barrels per day
(b/d) of liquid hydrocarbons. This includes conventional crude, synthetic crude, pentanes plus or condensate, and natural gas liquids—
ethane and liquefied petroleum gas (LPG). Crude oil production is
1.6 million b/d, excluding natural gas liquids. By comparison,
Canada's demand for crude as oil and equivalent increased 4.6 percent in 1979 to 1.87 million b/d (not including the natural gas liquid
demand).

Table 12.1 shows 1979 production and consumption by province.
It is most apparent that Alberta is the main energy-producing province, with about 1.4 million b/d of production and 274,000 b/d of
consumption. Canadian exports go almost entirely to the United States.
In 1973 they were about 1 million b/d, but are very small now,
amounting to about 150,000 b/d, most of that being heavy oil. Canada
sent another 140,000 b/d of oil to the United States in exchange for
other oil.

Canada actually is very close to self-sufficiency in liquid hydrocarbons, but unfortunately they are not all liquid hydrocarbons that
can be used, so Canada is importing 400,000 to 500,000 b/d.

Canada's big problem is that the established reserves of conventional crude oil have been falling since 1970 to the current level

of about 5.9 billion barrels. These figures do not include oil recoverable through exotic recovery methods. A few of these have been tried without very much success.

The ultimate potential for conventional crude oil in the provinces is estimated at approximately five times the proven reserves. It also should be pointed out that Alberta's conventional crude oil reserves are about 81 percent, down to about 11 years of production. Certainly the ultimate potential is much more optimistic, and as those familiar with oil estimates know, it may be overly optimistic or it may be unduly pessimistic. It is about 29 billion barrels, but a high level of investment and considerable time is going to be needed if these reserves are to be brought into production. As Table 12.2 shows, the locations are: British Columbia, a very small portion even of the ultimate; Alberta, with a potential that is about double the developed reserves; the main areas of interest are the Arctic Islands and the Arctic offshore; and then the Atlantic and Eastern Shelves, which are two major areas of activity at present.

Oil sands plus conventional reserves represent the largest potential. The designation "frontier areas" is obviously at this state quite unknown. There is no question that there is oil, but when it can

TABLE 12.2

Remaining Established Reserves and Ultimate Potential
of Conventional Crude Oil, 1979
(billions of barrels of crude oil)

Province	Developed	Potential
Alberta	4.80	9.00
Arctic*	—	11.00
Atlantic Provinces	—	5.00
British Columbia	0.18	0.30
Manitoba	0.04	0.20
Northwest Territories and Yukon	0.13	0.50
Ontario	0.01	0.08
Quebec	—	0.01
Saskatchewan	0.74	4.00
Total	5.90	30.09

*Not developed.
Source: Compiled by author.

TABLE 12.3

Marketable Natural Gas Remaining in Established Reserves in Canada, December 31, 1979 (Bcf)

Regions	Net Change during 1979	Nonassociated	Associated	Solution	Underground Storage	Remaining Reserves
MacKenzie Delta/ Beaufort Sea	-62.4	6,535.2	—*	—	—	6,535.2
Arctic Islands	3,230.0	14,113.0	—	—	—	14,113.0
Subtotal	3,167.6	20,648.2	—	—	—	20,648.2
Mainland Territories	-23.2	575.4	—	—	—	575.4
British Columbia	-390.5	7,010.7	274.3	75.1	—	7,360.1
Alberta	3,431.4	49,261.2	6,349.6	2,772.1	53.7	58,436.6
Saskatchewan	3.5	1,146.2	46.2	56.5	8.8	1,257.7
Ontario	-18.5	106.8	—	—	198.0	304.8
Other Eastern Canada	-0.4	10.5	1.3	—	—	11.8
Subtotal	3,002.3	58,110.8	6,671.4	2,903.7	261.0	67,946.4
Total Canada	6,170.0	78,759.0	6,671.4	2,903.7	261.0	88,594.6

* – Means none.

Source: Compiled by author.

242

be brought into commercial production and just how much it will be used is still subject to question.

Some Canadian experts think the goal of energy self-sufficiency for Canada is obtainable, but there is certainly a lot of work to be done to reach it in view of our declining reserves and our current crude oil deficit. Self-sufficiency is a necessary target for Canada in view of recent developments in crude oil markets.

On the natural gas side, the picture is really much brighter. Proven recoverable reserves, excluding frontier and unconventional supply, amount to about 75 trillion cubic feet (tcf) in 1979. The National Energy Board, in their hearings toward the end of 1979, considered these adequate to meet Canada's requirements for approximately 25 years. Export commitments, which remain at approximately 10 tcf, leave an exportable surplus of 4.2 tcf. That surplus has been committed to export, a good part of it through the prebuilding of the Foothills Pipeline, a link to the Alaskan gas pipeline. There have been quite a number of areas of disagreement between the Canadian government and the U.S. government and a number of commercial problems with that gas. The prebuild was supposed to be exactly what it says, a prebuild, and there is a lot of concern in Canada that it may turn out to be merely an oversized line used to export Alberta gas and that the Alaskan gas may be delayed considerably or may not even travel that route in view of the rapidly escalating costs.

Of the 4.2 tcf that has been licensed by the National Energy Board but still (1980) awaits government approval, almost 2.5 trillion goes to the Foothills Prebuild, about 700 billion cubic feet to California, and about 1.5 trillion on the eastern leg of the prebuild system.

Canada is now (1980) using about 1.65 tcf of gas per year and is exporting 1.01 tcf to the United States. This represents about 5 percent of U.S. natural gas requirements but in certain regions, of course, it represents a much higher percentage. In 1979, it was about 25 percent of the total natural gas used in the state, 65 percent in the Pacific Northwest, and 40 percent in Montana. Those areas receive about 70 percent of the total Canadian exports. Most of the balance is marketed in the Great Lakes area in Minnesota, Illinois, and Michigan. Small areas in upper New York State and Vermont are completely dependent on Canadian gas.

Canadian gas reserves shown in Table 12.3 are more optimistic than the proven reserves, although this graph actually shows the remaining established reserves. The approximately 70 tcf mentioned earlier represents the connected or readily connectable reserves. These include frontier reserves that have been proven but are not yet connected into the systems. There is great optimism over some of the areas. In fact, one of Canada's major explorers, Canadian hunter which discovered the Deep Basin in Alberta, is talking about 400 tcf in that basin. So far, the National Energy Board has included 1 tcf in its reserves. The figure is probably between the two.

At 1980's prices, there probably are quite a few trillion cubic feet that can be taken from that area. The success ratio in finding gas is quite good, but it is still too early to know how durable it will be.

The prospects for petroleum from frontier reserves show promise, but the main source of Canadian petroleum will be the large bitumen deposits found almost entirely in Alberta. The reserves are estimated to be close to 1 trillion barrels of oil, but available technology would allow recovery of only 8 percent. If one considers economics as well as technology, this figure is probably about 3 percent. In sum, on the order of 30 billion to 40 billion barrels of oil could be extracted from the sands at today's (1980) economics and technology.

There is considerable exploration activity in the frontier regions. The rig count in 1980 was about 500 and, to the end of June 1980, about 4,000 wells had been drilled. However, those are largely development wells, many of them shallow wells, which is certainly not indicative of the drilling in the frontier regions.

There is some natural gas on Sable Island, although it is not yet commercial. East Shelf has two discoveries, Hibernia 35 and 15, while Labrador Shelf and Baffin Island have nothing really commercial, but lots of hope is offered. There are also the far Arctic Islands, the Beaufort Sea, Norman Wells, one of the discovery fields, and still quite a bit of activity in Alberta and British Columbia, with the Deep Basin area that has between 1 and 400 tcf.

Of Alberta's bitumen reserves, the Cold Lake reserves, containing 159 billion barrels in place, are a little more easily producible by subsurface means. The Wabasca, Athabasca, and Peace River reserves, containing 24,720 and 64 billion barrels are much finer sands, and are much more difficult to manage, but do represent, certainly, tremendous potential and hope for the future.

There are two synthetic oil plants operating in the Alberta Tar Sands. The first commercial facility came into production in 1967 when Great Canadian Oil Sands—then a Sun affiliate, now Suncor—constructed a 45,000 b/d plant near Fort McMurray. This operation involves open pit mining of raw oil sands, separation of the bitumen by a hot water process, and upgrading to produce sulfur-free, high gravity synthetic crude oil. It is currently being expanded to 65,000 b/d.

In 1978, the $2.4 billion Syncrude project came on stream with a designed capacity of a 125,000 b/d of synthetic oil. It uses fluid coking, as compared to delayed coking at the smaller Suncor plant. Giant draglines are used instead of bucket wheels to mine the sand. The project, now producing close to 100,000 b/d, is expected to reach its planned capacity in 1982. However, there are plans to expand it.

The technology of these operations is constantly breaking new ground. The Syncrude project experienced significant start-up difficulties, much of which related to the coking system. The fluid cokers used to upgrade the bitumen are based on proven technology, but they are certainly a lot larger than anything else in use, and that is responsible for quite a few of their start-up and initial running problems.

Alberta is at the forefront of recovery technology for oil sand processes. Through research, much of it funded by the Alberta Oil Sands Technology Research Authority (AOSTRA), Alberta is pressing ahead to find methods to improve recovery and to develop these deposits, particularly those deposits that cannot be mined by surface methods. Out of $200 million committed to AOSTRA, about $60 million has been spent since 1976.

Three additional projects have been proposed for Alberta's oil sands. A group of companies headed by Shell Oil Canada (Alsands) has applied to the Alberta government to construct a third oil sands mining plant of 140,000 b/d capacity for completion in 1986 at an estimated cost of $6.8 billion.

Esso Resources Canada Limited has applied to the Alberta government for a permit to construct a steam-injection project in the Cold Lake area with designed capacity of 140,000 b/d of synthetic oil. This project would start production in 1987 and will cost $7 billion.

A fourth oil sand mining plant proposed by Alberta Gas Trunkline/Petro-Canada is in the formative stages. This plant would be of a size similar to the Syncrude and Alsands plants, but would not come on stream until 1990. Total costs are estimated very roughly at $10 billion.

In the very short term, there is some potential to improve energy supply security by substituting natural gas for fuel oil, by upgrading refining to reduce heavy fuel oil, and by utilizing more LPG in petrochemicals and motor fuel for road machinery but not for automobiles. These steps could save 300,000 to 500,000 b/d of petroleum imports. The approach is not to attempt to sell natural gas in direct competition with the heavy fuel oil for it is felt that the oil is too valuable and should be upgraded. Canada should not be producing nearly as much heavy fuel oil as it is.

These are the technically feasible solutions to Canada's energy problems, but the question is why more of them are not being implemented. Canada's energy situation is complicated by two political issues: ownership and pricing.

Under the British North America (BNA) Act, which is the constitution, the ownership of resources and the right to develop and sell these resources rests clearly with the provincial governments. This includes disposition by lease or sale, collection of royalties, good production and conservation practices, and maintenance of reserves adequate to serve provincial needs. But the British North America Act gives to the federal government jurisdiction over "works or undertakings connecting the province with any other of the provinces." Interprovincial pipelines clearly fall under federal jurisdiction. The Canadian government can regulate interprovincial and international trade (including price), and its taxing powers are almost unlimited, subject, of course, to voter reaction.

Over the past six years Canada has had to resolve conflicting claims of jurisdiction concerning energy and, in particular, the development of natural resources and the pricing of oil and natural gas. The conflict is evident. The BNA Act gave the province control over resources in the ground, and the right to sell in the province, but also gave the federal government authority over interprovincial trade and export from the country. The constitution was probably set up purposely to be vague because agreement could not be reached on specifics when it was written.

It is the second issue mentioned previously, pricing, that has really brought the ownership issue to the fore. In 1974, when the international prices rose so dramatically, the Alberta and Canadian governments agreed to stage price increases at a rate of $1 per barrel every six months until world levels were reached. Obviously, no one at that time foresaw the large price increases that would be instituted by other producing nations. The current (1980) Canadian average domestic wellhead price, under the agreement, in Canadian dollars, is $14.75 per barrel, which translates to about $16 per barrel at the refiners' gates in eastern Canada. This, of course, can be compared with $33 per barrel average landed cost for imports in the same market, even before the last set (1979) of OPEC price increases are brought into account. Canada prices natural gas at about 85 percent of the crude oil Btu-equivalent value.

In conclusion, it should be emphasized that Canada has tremendous potential. A surplus of natural gas now exists that may become larger, and with great effort, the country can become self-sufficient in crude oil. However, if this is to take place, all resources must be developed and used to their greatest potential. The solution of the political questions will strain the very fabric of the country. There is little doubt that a solution will be achieved and that Canada will continue energy-secure well into the future.

13

CALIFORNIA OIL SUPPLIES

Dennis Eoff

This chapter explores California's potential for increasing its own oil production and some of the constraints that might be involved. California is currently the fourth largest producer of oil in the United States, with production of 347 million barrels in 1978, or about 950,000 barrels a day (b/d). Production in 1980 was slightly over that. This accounts for about 40 percent of California's total use, with the remainder coming from Alaska and foreign sources, primarily Indonesia.

Production peaked in California in 1968 at a little over 1 million b/d and declined until Elk Hills was opened in 1976. Since then, the production has climbed slightly as Elk Hills' thermally enhanced oil production has increased.

According to the California Division of Oil and Gas (DOG), California's reserves, as of the end of 1978, were 4.2 billion barrels. They have not released their official reserves for 1979, but unofficially they are in the neighborhood of 5.2 billion barrels. The increase is not due to any new discoveries. It is rather a reevaluation of known resources due to price decontrol of heavy oil. The most current American Petroleum Institute (API) estimate is about 3.6 billion barrels. In addition to this, API has added 1.6 billion barrels of what is called "indicated" reserves. API is not calling those proven reserves, but once again the new category is primarily related to the economic viability of producing heavy oil. These "indicated" reserves are for the most part located in Kern County.

Two kinds of production in California probably have the most promise for increases: thermally enhanced oil production and offshore outer continental shelf production (OCS). Some other areas that might show increased production are Elk Hills, some new—and

247

probably minor—discoveries onshore, and old fields, reworked in response to higher prices.

How much oil can be recovered by thermal enhancement depends on how much oil California has in place. Estimates range between 35 billion to 60 billion barrels. According to the DOG estimates made in August 1979, there are approximately 4.2 billion barrels of proven reserves in California. Of this, about 2 billion is 16° API gravity or less, which is where most of the thermal enhancement would take place.

In 1978 California produced about 400,000 barrels of oil that was 16° API or less, about 240,000 of which was produced by thermal enhancement or steam injection. Most heavy oil is in Kern County in the southern San Joaquin Valley, but there also is some steam production in the Los Angeles Basin and the coastal area at the San Ardo field.

The estimates of how much oil will eventually be produced from this 35 billion to 60 billion barrels of oil in place range from 7 billion to 12 billion barrels. However, even after this year's (1980) price increases, DOG is apparently going to add only about 1 billion barrels to last year's proven reserve estimates, for a total of 5.2 billion barrels.

In a 1978 report to Congress, the Energy Information Agency estimated potential additional production by 1990 as high as 1.51 million b/d. More recent estimates are in the neighborhood of 100,000 to 530,000 barrels. The difference in the figures probably has to do with the problems of producing heavy oil. The resources may support the higher end estimates (the million barrels), but the problems of capital intensity for developing the oil precludes production at those rates. The number of production and service wells is quite large. The cost and amount of needed refinery retrofitting and the amount of steam generation and pollution control equipment are also constraints. There are also the negotiations regarding air quality that have been going on for a number of years. In the last year the Air Resources Board and the producers seem to have reached an agreement with regard to NO_x and SO_x emissions. The producers are not entirely happy, but at least feel that they can now make investments and increase production. There is still some question as to the amount of new production the new SO_x rule will allow. There is also a question about particulate emissions.

Beyond that, the problem of producing more heavy oil is the problem of what to do with it. A number of California refineries cannot process it because of the high content of metals and sulfur and the gravity of the oil. Even when they can process the heavy oil, the resulting product slate is low on the light products such as gasoline and very high on heavy—and less valuable—residual fuel oil. The residual oil will probably continue to decline in value relative to gasoline and to the Fuels Use Act and air quality standards.

The Bonner and Moore study, a cooperative effort between the state government and oil companies, took a detailed look at what to do with the heavy oil. The California Energy Commission is also initiating two further studies. The first examines the constraints to heavy oil production and what might be done to alleviate them. The second examines various options for dealing with the heavy oil and the expected surplus of residual and how the preferred option should be implemented.

Offshore oil production can be divided into two areas: state lands that are within three miles offshore, and beyond that, the federal Outer Continental Shelf (OCS). Production in state lands has been declining even though the Wilmington field, which is the largest in the state, produces 37 percent of its production from offshore areas. Production from this field, other offshore areas, and federal OCS areas are all decreasing.

There are a number of plans for additional production platforms in the federal OCS area. The Hondo field is apparently ready to produce approximately 40,000 b/d. If you total up all the production estimates for known discoveries in the federal OCS, the maximum production is a little bit less than 150,000 b/d. But all of these platforms will not reach their maximum production at the same time. Over the next five to ten years, which is the probable development time for even new discoveries, the maximum additional production probably is not going to exceed 150,000 b/d.

There is obviously a lot of resistance and problems with leasing and developing these fields due to environmental considerations and there is an extremely long lead time in developing anything that is out there. Whatever oil is available will be produced, but over a longer period of time than some of the estimates from a few years ago. Consequently, the maximum rates will probably be lower than if all the fields were developed and produced at the same time.

The Elk Hills Reserve originally was scheduled to produce over 300,000 b/d, possibly even 350,000 b/d. The most recent (1980) federal government forecast is 210,000 b/d, compared to the 155,000 b/d that it is currently producing. Estimates were reduced because processing facilities were not started on time, there has been some litigation, and less oil has been found than expected at deeper levels.

There is also another problem that could affect Elk Hills production: the plan recently passed for filling the Strategic Petroleum Reserve (SPR). That legislation requires that at least 100,000 barrels of oil be added to the SPR per day. If other supplies are insufficient, the fill rate will be maintained by transfers from Elk Hills. Although this probably will not affect the amount of oil available in California, there might be some dislocation to refiners.

California has been well explored onshore so large new discoveries are not expected. The largest field that has been discovered in the last five or six years is the Yolume field, which is currently (1980) producing 24,000 b/d. This is not a large amount, but one cannot expect any larger discovery than that onshore.

In summary, our projections show an expected production level of about 1.1 million b/d in 1985 and 1990, and 950,000 b/d in the year 2000. The maximum levels are consistent with the high ends of the ranges just discussed.

The benefits of an increase in California production are well-known. It would reduce the demand for foreign crude, help the national balance of payments, be a stimulus to California's economy, and it might reduce our vulnerability to supply offsets.

There are two costs that should be weighed against the benefits. First, California may wind up shipping other types of oil to the rest of the country: in effect, importing somebody else's pollution. The second point is whether or not maximizing California production is a "drain California" policy. If the SPR is a good idea for the country because it reduces the country's vulnerability to imports, it might be a good idea for California. As California increases production and depletes reserves it may become more vulnerable to imports. California's heavy oil reserves could be its own state "strategic reserve."

VI
Strategic Policies to Counteract the Dependence Dilemma

INTRODUCTION

The authors in the preceding chapters outline the basis for the continuing dependence of the United States on foreign oil. There is most evident agreement that a supply disruption of unknown size probably will occur sometime in the 1980s. Even though some public officials and spokespersons from the private sector periodically proclaim that energy problems are manageable, reports continue to appear that concur with the views expressed in the previous chapters. For example, the staff of the United States Senate Committee on Energy and Natural Resources concluded in December 1980, that the United States would still be importing oil at the end of the century, no matter what policies it pursues.[1]

The report also argued that energy problems involve broad questions of national security related to the economics of western allies, the explosive Middle East, and the Soviet Union. In effect, the United States can never be independent because no matter how low the level of oil imports, it remains vulnerable to a cutoff of supplies because the nation's allies are vulnerable and will never be energy independent. In view of this position, the report called upon the United States "to place less emphasis on reducing imports and give far more attention and financial resources to emergency preparedness and stockpiling."[2]

The contributors in Part VI focus their attention on a number of the issues and controversies surrounding energy security. Although they disagree over questions such as private versus public stockpiling, or a mixture of the two, and a variety of other aspects of such a program, there is general agreement that some type of oil reserve is highly desirable.

It also is most evident that even if policymakers and experts agree on the need for a reserve supply of oil, there will be much disagreement over how to establish, maintain, and utilize such a reserve. Indeed, the political complexity of the issue, the variety of available options and past experience with stockpiling, all of which are outlined clearly in this part, suggest that developing a reserve will sorely strain the capacity of governments at all levels and the private sector to develop and implement such a reserve capacity to serve as at least a partial barrier from the drastic economic and political impact of consequences of a substantial world oil shortfall.

NOTES

1. Christopher Madison, "A Note of Dissonance," The National Journal, December 6, 1980, p. 2089.
2. Ibid.

14

ENERGY AND SECURITY: FEDERAL AND STATE POLICY IN A CHANGING INTERNATIONAL ENVIRONMENT

David A. Deese

World oil supplies have been seriously disrupted five times since the late 1950s. These disruptions are occurring at an accelerating rate, with three during the past seven years. Structural change in world oil markets and political change in major oil-producing countries in the third world make future interruptions highly unlikely. As of 1980, it remained entirely possible that the war between Iran and Iraq could escalate and temporarily stop the entire 20 million barrels per day (b/d) of crude oil and petroleum product exports that passed through the Straits of Hormuz.

What would such an event mean for other countries and the United States? Before the Iran-Iraq war, Japan imported over 3.6 million b/d of crude oil and petroleum products through the Straits of Hormuz, France imported almost 1.8 million b/d, and the United States received almost 2 million b/d, or 25 to 30 percent of all its imports. Even if the International Energy Agency (IEA) sharing formula worked as designed—and this is open to serious question for such a large disruption—Japan would face a total supply reduction of almost 2.5 million b/d, France over 1 million b/d, and the United States about 4.5 million b/d (based on a 10 percent reduction in consumption in 1979 of about 1.7 million b/d and the use of 2.8 million b/d in emergency stocks). For the United States, this means the entire Strategic Petroleum Reserve would be depleted in little over one month—a scenario that almost guarantees that U.S. leaders would not be willing to use the reserve, and thus would be unable to meet their legal, moral, and political commitment to the IEA.*

---

*Private stockpiles have been built up to over 250 million barrels of crude oil and petroleum products that could be used without

254

While the specific results of such a disruption are impossible to predict, certain events are highly likely. As oil became scarce on international markets, companies and governments would begin to panic. Spot market prices would rise rapidly. Product prices in each country would soon reflect this increase to the extent that they were not controlled. In countries with price controls, such as the United States, gasoline lines would form and governments would resort to the cumbersome task of allocating gasoline to retailers on the basis of historical use patterns. As the crisis deepened, the inconvenience of lines would give way to more threatening social and economic disruptions, including factory closings and possible spot heating fuel shortages.

If individual states were prepared, despite the federal failings, private and public oil stocks and other measures such as sticker plans or employer-based car pooling schemes could help ease shortages and reduce panic. Yet without strong federal preparations, states cannot insulate themselves from disruption. Increased supplies or lower prices in one state would quickly be balanced out among neighboring states, leaving a well-prepared state only somewhat better off.

Yet even marginal reductions in social and economic disruption through prudent preparations could have additional benefits in a crucial state like California. If panic is minimal in California, important emergency preparations such as diverting Alaskan oil or imports from Indonesia and Mexico to crude-starved sections of the United States could be much easier to execute. This could help ease pressure on international, national, and regional markets and reduce the probability of serious political strains among the Western consuming nations.

DEFINING THE ENERGY AND SECURITY THREAT

The energy and security problem confronting oil-importing nations has three components. First, as described, oil supply interruptions of all sizes can disrupt normal social and economic activity. The first signs of such physical shortages in the United States in the past have been gasoline lines. Second, sharp increases in crude oil and product prices that accompany even small oil supply disruptions impose heavy economic costs. These economic problems aggravate political relations among and within oil-importing nations.

---

disrupting the U.S. distribution system. This could, in principle, provide another 80 days of stockpile use at 2.8 million b/d, but there is no means to force the oil companies to use this oil.

The final energy and security threat is political coercion imposed on an oil-importing nation to adopt positions favorable to oil exporters. This coercion takes the form of support for the Palestinians and opposition to Israeli stances, trade and investment concessions, and the transfer of sophisticated technology, including sensitive components and materials from the nuclear fuel cycle.

How likely are further oil supply disruptions in the 1980s? Analysts at the Department of Energy (DOE) have synthesized commonly held expert opinions on this question into a set of hypothetical probabilities for each of three views of the world. All three are possible, with World View 1 the most optimistic and World View 3 the most pessimistic about the likelihood of future disruptions (see Table 14.1).

These disruptions roughly represent, for example, all oil produced by Iraq (3 million b/d) before the current war, by Saudi Arabia (10 million b/d), and by all producers in the Persian Gulf (20 million b/d). At the level of 3 million b/d, World View 1's estimate of a 50 percent probability for disruption may already be overly optimistic. Given the disruption of almost 4 million b/d caused by the Iran/Iraq war, these world views might all shift to somewhat higher probabilities for the 20 million b/d disruption. Yet even using this version of World View 2, which may represent the greatest weight of expert opinion, the probability of 30 percent for a 10 million b/d disruption in the 1980s demands serious attention.

Under current federal policy, even a prolonged disruption at the 3 million to 5 million b/d level could cause severe problems. Since no effective stockpiling, allocation, or demand restraint measures are available, the only option is to resort to price controls and gasoline rationing. During the several months that it takes to get in place the burdensome machinery for coupon rationing, gasoline shortages would be allocated by long lines.

What is the energy and security threat to California? California has a particularly important role in providing both the United States and its own energy security. It has direct and special relationships with Mexico, Canada, Indonesia, and, to some extent, Japan. Despite its increasing access to Alaskan oil, California can help assure the continuation of low-sulfur crude exports to the United States from Indonesia and the expansion of gas and crude exports from Canada, Mexico, and even Venezuela. U.S. federal government efforts in these areas have generally been so poor that the states and regions of the country have not only the opportunity but also the duty to take further initiatives on their own.

As the fourth largest oil-producing state and the single most important state on various measures, California can exert constructive pressure on the federal government and other state governments

TABLE 14.1

Hypothetical Probabilities of Oil Supply Disruptions

| World Views | Size of One-Year Disruption | | |
	3 mbd	10 mbd	20 mbd
1	50%	10%	5%
2	75	30	5
3	95	50	25

Note: Percentage probability for occurrence at least once in decade.
Source: Tom Neville et al., The Energy Problem: Costs and Policy Options (Washington, D.C.: Department of Energy, 1980).

through its congressional delegation, governor, state legislature, and other public and private organizations. Leadership is required in restraining state initiatives that would be counterproductive for broader U.S. interests, and for prodding the federal government to take effective steps. The federal government can be influenced either directly with pressure to take specific actions or indirectly by setting an example with effective action at the state level. One crucial example that combines elements of both types of action would be cooperation in the diversion to Japan of some Alaskan oil normally destined for California during a severe supply disruption.

California can also be an energy laboratory for the United States and other countries. Data is scarce or lacking on the effects of many important energy security measures. Past and future initiatives in stockpiling, emergency allocations, controlling consumer panic, emergency demand restraint, and emergency supply can serve as innovations that simultaneously increase both the United States' and California's energy security.

PRINCIPLES FOR ACTION

Be prepared, but do not key plans to specific predictions. It is absolutely crucial to have preparations in place before problems arise, yet highly detailed plans may turn out to be useless. The energy and security threats defined above cannot be predicted with accuracy. All three oil supply disruptions during the past decade caught many by surprise and the last two, in 1979 and 1980, caught almost everyone off guard. Furthermore, the specific causes, depths and durations of these disruptions are unclear or unknown until events are already underway or even over.

Despite this ambiguity, it is extremely important to prepare responses in phases according to the size of oil supply disruptions. The keying of specific sets of measures to three levels of disruptions (possibly of up to 5 million b/d, between 5 billion to 10 million b/d, and over 10 million b/d for one year in the world market) is useful, but most important is maintaining overall flexibility in response systems. Some measures may be useful only for small disruptions, while others are necessary and appropriate only for large ones.

Be prepared to act in a specific time sequence. Measures such as restraining demand and selling stockpiled crude oil or products may be extremely effective if implemented in advance of the actual supply disruption. Since lags are built into the oil supply system, especially in the long tanker travel times at sea, it may frequently be possible to take small but highly effective preventive measures even before the crisis strikes. Once such action is taken, national and state governments can watch and wait to take further action as required. Some measures, such as gasoline rationing by coupon or emergency supply through increased natural gas or coal production, may not be available until the later stages of the crisis.

Use the market whenever feasible. Market imperfections and political realities restrict the effectiveness of the market under some circumstances, but it is also difficult for regulatory actions to be efficient and equitable. Careful and detailed attention should be given to each government intrusion into the market because complex problems of political economy are rarely solved satisfactorily by direct government intervention.

If monitored carefully, allocation by price may frequently be the quickest and most equitable measure available. It may be during emergencies that the market system is both the most needed economically and the most difficult to maintain politically. Yet market mechanisms may be an option available to governments if they are combined with the whole array of other energy security measures that can be prepared in advance.

Be ready to capture excess profits and get rebates to consumers. The windfall profits tax is an invaluable part of energy emergency preparations. This may be the only way to capture the enormous rents created during disrupted conditions and still allow the market to function. Rebates are also crucial. To maintain equity, to control inflation, to keep consumers' credibility, and to allow the government to concentrate on the most important issues, it is essential to rebate most of the windfall profits to consumers or households. In order to induce changes in consumer behavior and to allow the government to handle large financial transfers, some lag—perhaps 30 to 60 days—in providing rebates will be necessary.

Enact measures automatically. The United States must escape the current situation where it would automatically revert to the entire system of price controls and some gasoline rationing system by default. For many types of emergencies, doing nothing would be highly preferable to our current lack of preparedness. A sequence of self-executing measures should be in place so that governments do not have to act in haste during an emergency. The time for careful action is well before the intense social, economic, and political pressures imposed by emergencies.

ENERGY SECURITY MEASURES

States should carefully investigate the entire spectrum of possible energy security measures from increased Indonesian imports to accelerated production of electricity at existing power plants. These measures can fit into any of three categories by origin:

Federal	Federal and State	State and Local
Strategic Petroleum Reserve	Private stockpile	State petroleum reserve
IEA Sharing System	Emergency taxes	Gasoline sales sticker
Surge production arrangements with other countries	Speed limits	plans or minimum
	Public sector limits on building	purchase rules
Diplomacy	temperature or commuting	Emergency information system activation
Military actions	Compressed work week	Waiving of restrictions on utilities
	Diversity of supplies	

The listed measures are only examples. There is no attempt to be comprehensive. Some can only be originated at the federal level or with federal cooperation because states lack the authority to act—for instance, voting to trigger the IEA sharing system or taking military action—or because individual state action might be ineffective—for instance, releasing state or private stockpiles or imposing an emergency tax on gasoline. In each case, the states should be urged to proceed as far as feasible without interfering with federal action.

To date (1980), coordinated federal and state action has only been attempted in emergency conservation. Yet all the efforts leading up to and including the Standby Federal Emergency Energy Conservation Plan of 1980 leave much to be done. This federal-state conservation system leaves considerable doubt as to what incentives, especially funding, the states have to act; what real room there is for

the states to act without having federal action takeover; how equitably
the federal government can set individual state conservation targets;
and how effectively federal conservation standards are enforced by
the states. But while improvements are sought in this area, serious
action is possible on many other fronts.

First, California should seriously consider maintaining Indonesian imports, if not on a regular basis then perhaps during emergencies, in order to send Alaskan oil further east in the United States.
About 20 percent of California's oil consumption is met by oil imports
from Indonesia. More important, all shifts that occur in the level of
California's imports from Indonesia should be carefully monitored
and coordinated with broader U.S. energy security objectives.[*] All
possible technical and management assistance should be offered to
Indonesia in its domestic energy policies. Without important policy
changes, especially in product pricing, all of its production could be
consumed internally by 1990—95.

Second, external sources of energy supplies should be diversified as much as possible. This includes arrangements for both normal
and disrupted conditions. When it is activated, the IEA system offsets
the effects of high concentration of oil import sources. But it remains
important for California to diversify its import sources to cover conditions below the IEA trigger, before the IEA system comes into effect, and beyond the capability of the IEA system. But California
poses a special case since about 40 percent of its consumption is
satisfied with Alaskan oil. Although this source is threatened both by
accidents such as fires in producing and collection areas and by sub-
marine attacks on tankers during war, Alaskan oil is probably relatively secure when compared with any other external source. Diversification away from Alaskan oil may therefore reduce the magnitude
of the maximum feasible supply disruption. Yet, depending on the
replacement source, such diversification could also increase the
probability of disruptions.

Diversification must be monitored very carefully in two further
respects. As recommended above, there should be extreme caution
in maintaining excellent relations with the countries that now provide
or may provide oil imports. Additionally, the types of crude and
products available must match local refinery capabilities. If, for
example, low-sulfur crude from Indonesia is lost during an emergency,

---

[*]Since Indonesian crude is used primarily in power plants, reductions in power plant usage involve turning away Indonesian imports—
an action that would clearly violate broader U.S. interests.

replacement crude or refinery capacity must be found. As the world market gradually shifts to heavier crudes, the match of crude to required products and refinery structure will become an increasingly important element of energy security.

Third, California should do all it can to help maintain the capacity to divert Alaskan oil to Japan during a major disruption. The U.S. public understanding of problems associated with this step could be severe. As a local citizen, it may be especially difficult to accept the sharing of U.S. resources during a crisis. Yet it is essential to begin the process now of alerting the public to the overriding need to help keep Japanese companies out of the spot market and to maintain cohesion among the allies.

Fourth, California should carefully investigate the possibilities for public and private oil stockpiles. With or without effective federal action, it may be feasible to create state reserves, especially of petroleum products, and to encourage the holding of private stocks. The entire spectrum of private sector stockpile options should be considered, from offering tax incentives, to providing storage facilities, to regulatory requirements for importers and refiners to maintain at least specified minimum levels. European precedents should be carefully examined. Intensive consultations should begin with the federal government on the possibilities for state and regional stockpiles.

Building these revenues, however, is only a start. Stockpile management involves difficult questions of how and when to release stocks, what price to sell them at, and how to rebuild them. There is a further complex question of how much of each product to hold in reserve. Each step taken into these areas by a state government must be supported by reliable information flows. State governments should also consider the possibility of building other types of reserves, including natural gas and any other possibly applicable fuels.

The fifth measure includes shut-in, or surge, production capacity for crude or natural gas. Shut-in production capacity in other countries proved to be crucial during the supply disruptions of 1979 and 1980. While it can only be of use during a directed embargo if the producers split into feuding blocs, under chaotic interruptions of world oil supply it is at least as important as stockpiles as an energy security measure.

What role can California play in this question? All possible avenues should be evaluated for creating surge production capacity locally, perhaps at the Elk Hills Naval Petroleum Reserve, in Alaska, and in other countries. California could, in close coordination with DOE, quietly pursue the possibilities of emergency crude oil supplies from Mexico and natural gas supplies from Canada.

Closely related to this measure of surge production capacity are all the other possible increases in energy supply, including

wheeling of electrical power that would affect California. Some short-
term supply increases can be produced by eliminating maintenance
plans for electrical power plants and running all facilities at full
capacity. Yet many of the supply options require months of lead time
to reach full effectiveness.

The sixth category covers emergency conservation. There has
been considerable action with few tangible results in this area. Despite the uneven quality of state conservation plans and programs,
measures at the state and local level are particularly important.
Consumer panic may best be avoided by state action, and most emergency conservation actions must be initiated or enforced at the state
level. This applies particularly to utility conservation programs,
which hold considerable promise for both normal and disturbed market conditions. The option for state and local government use of
taxes, especially surcharges, under the Emergency Energy Conservation Plan of 1980, deserves very serious attention. The recent use
by Los Angeles of a tax on water usage above a specified level should
be assessed for possible applicability to energy tax policy.

The final category of emergency measures is the displacement
of petroleum use, including the retrofitting of the refinery industry
to provide badly needed flexibility under both normal and disrupted
market conditions. A number of important advantages accrue from
displacing petroleum use. Reduced oil use decreases the deadweight
economic losses from the enormous transfers of wealth caused by
sharp price increases. It also reduces strains on the world oil market,
allowing the level of unused production capacity worldwide to increase
and possible reducing the reliance on particularly unstable sources
of supply. In the case of California, however, this question is more
complex since Indonesian exports are especially important to broader
U.S. energy needs. The reduction of petroleum use is also a key
focus of conservation efforts, our commitments to the IEA and the
allies, and the image of U.S. leadership worldwide.

These general advantages of oil import reductions have, however,
probably been overemphasized. They are necessary but not nearly
sufficient for U.S. energy and security interests. Import reduction
remains the centerpiece of both domestic and foreign policies. Enormous financial and political resources are spent on synthetic fuels,
the proposed Energy Mobilization Board, and unilateral and multilateral oil import ceilings. Some still carry this measure to the absurd-
ity of urging complete U.S. self-sufficiency in energy supplies in the
1980s or 1990s. National, state, and local governments must work
together to dispel the notion that oil import reduction is our supreme
energy policy objective. It is absolutely essential for all the various
U.S. publics to understand that import reduction is no more or less
than one small step in the right direction.

Given the urgency of the energy and security threats and the dramatic failure of federal responses, increased regional and state action is essential. A division of efforts into supply-related contingency preparations in Washington, D.C., and demand restraint preparations in the states is no longer acceptable. The federal-state program in demand restraint is still plagued with severe problems, leaving serious doubt as to its effectiveness.

Under these conditions, state and regional coalitions should consider the entire range of energy emergency measures outlined above. Many steps require coordinated federal-state or state-local efforts, and all steps demand caution in not violating broader U.S. national or foreign policy interests in energy and security. But state roles in energy emergency preparedness include not only direct action at home but also urging action in Washington, D.C., and offering all possible data, experience, and examples.

15

INTERNATIONAL SHARING AGREEMENTS AND DOE's OIL CONTINGENCY PLANNING

William B. Taylor, Jr.

Due to space constraints, the International Energy Agency (IEA) and its sharing agreements will be somewhat slighted in this chapter. As has been discussed elsewhere, the world oil market will do a lot of what the sharing agreements are intended to do. Indeed, the IEA was set up about 1974 to deal with an embargo, and there have been discussions earlier about how successful producing nations can be in effecting a targeted embargo. It is only recently that the United States has become vulnerable to such a targeted embargo, so it is not clear what it has been doing with the IEA over the past six years. However, it may be that the IEA agreements may have gotten in the way had the United States had an interruption up to now. That may not now be the case, but let us go into the contingency planning being done by the Department of Energy (DOE).

On the supply side, stocks, either public or private, seem to be the only way of reducing the cost of an interruption, and clearly you have to use the stocks. It is not just possession, but the ability to use them in an emergency that is critically important. The administration, DOE, and Congress have not come to any agreement on mechanisms for drawdown. An auction of Strategic Petroleum Reserve (SPR) oil appears to have many advantages. However, there are obvious institutional kinds of problems for getting that done.

The sounds coming from Washinton, D.C., notwithstanding, DOE thinks it is critically important to buy a lot of oil very quickly while the market is slack and put it in storage. However, that is not exactly the message that is emerging when the administration takes money away, and Congress takes money away. Still, the DOE thinks it is very, very important to purchase oil—it is the only thing that can be done to offset or to mitigate the cost of an interruption.

The 100,000 barrel per day (b/d) minimum that Congress has set may—hopefully will—prod DOE into taking some more immediate steps. However, it is not enough just to have large volumes of oil in the ground in Louisiana and Texas. You have to be able to draw it down. You have to have the will and the means to draw it down. One way to avoid some problems may be to put a padlock on the first 300 million barrels or so and not touch it; then be willing to use the rest—assuming the SPR ever gets above 300 million barrels in it. If there were some automatic mechanism by which the "tactical reserve" could be drawn down, the 300 million barrels would be a "strategic reserve" that could be used to provide fuel for the armed forces or whatever. But even if there were some automatic mechanism just to offset the problem of deciding to draw down—a kind of psychological command problem—several questions would remain. How do you draw down? What mechanisms do you use to allocate this spare oil? If you leave it to DOE as directed by Congress, it is likely to pass it out "equitably." It will use some kind of buy/sell system or it will use some kind of allocation mechanism, so that every refiner, large or small, whether he can make gasoline, home heating oil, or only make residual fuel, will get the same amount—the same proportionate amount. It is not clear that this is the best way to use a resource during a time when that resource is very scarce. A better way may be to sell the SPR oil or to auction it, in which case the refiners who can do the most with it would presumably bid the highest and obtain those SPR oil supplies. The government then has sales revenues to do with what it will.

Leaving the supply side at that, it is clear that the SPR is critically important. Unfortunately, it is not a solution nor an option that we have in our contingency planning arsenal today. We have 92 million barrels in the reserve, but that may not last very long. That is 1 million b/d for 90 days.

Going to the demand side, during a supply disruption, the problem is that you have a limited quantity of available supplies. Consumption will clearly equal that limited quantity of supplies, but the question is how to ration them. Do you do anything to demand? At a conference earlier this spring (1980), Congressman Clarence J. Brown from Ohio had a great one-liner. He said that rationing is like sex education—if you don't do it, it will get done anyway. The point is, if the government does not step in and allocate the available supplies, they will be allocated. So the question becomes, "What kinds of mechanisms should DOE or the administration or the government develop and have ready to implement to allocate supplies?" Some interesting and useful criteria would be the use of market mechanisms and windfall profits taxes. There are criteria in addition to efficiency: the macro effect on the U.S. economy, fairness, and

practical problems. These we should bear in mind when we are evaluating different methods of allocating oil. Without detailing the intri-
cacies of each, there are four representative methods of allocating
oil supplies during an interruption.

The first, and the one that is at least somewhat familiar and
has been around since 1973, is to use the same actions taken in
1973—74. This regulatory option actually began in 1971 with wage
and price controls. Price controls were placed on domestic crude
oil. The government then froze or put limitations on the margins of
downstream operators (refiners, jobbers, and retailers) in an attempt
to hold down the prices and avoid windfall profits. If you strip the
price mechanism of its ability to allocate these scarce oil supplies,
then you have to have some other way of allocating these supplies.
This was done according to historical usage. The downstream operators are all required under the regulations to supply the same people
that they did in the past in the same proportion. However, this plan
has had its problems. One involves the problems associated with
the price control and allocation program. It is not flexible. It has
difficulties and causes lines. If prices are controlled below the market price, there has to be some way to allocate supplies other than
by price; lines seem to do it in this regulatory case. Also, price
control and allocation guarantee to some oil companies or suppliers
some supplies that they might not otherwise get. Some oil firms
stay in business that might not otherwise.

One thing that should be pointed out about lines is that you do
not seem to get any conservation until the lines start to form. This
could be broadened a bit to note that you do not get conservation until
the price of gasoline, broadly defined, increases. The point is that
even though you are paying in dollars a controlled price for gasoline,
you are still paying a high price. In fact, you are paying a high price
by waiting in line. The true price, the effective price, is the real
cause of conservation.

The second method of allocating supplies is the coupon gasoline
rationing plan for which DOE has worked long and hard ever since
1974. After several iterations, several administrations' efforts, Congress took a look at the next-to-last version in 1979 and said, "No,
we don't like it, do it over again." DOE has recently (1980) submitted
it again to Congress for its review and approval, and the law under
which they are reviewing it makes it more difficult for them to reject
it.

For coupon gasoline rationing to work for any lenth of time (and
this may not be well perceived or widely understood—it certainly was
not within DOE) you probably need to have price controls on crude
oil, and also price or margin controls on, and allocations of, downstream products other than gasoline. Why is that true? Well, in

order to ration, you must be sure you need to ration. You have to create a shortage in some sense. One way to create a shortage, as discussed, is to put on price controls below the market clearing price. That is guaranteed to do it. Then, but only then, can you ration. If the world market raises the world oil price, and presumably the domestic price of crude oil, to market clearing levels, then there will not be an excess demand for gasoline, and the world market will, in fact, eliminate the need for rationing. That is the first thing to understand and the first thing DOE is going to have to do is reimpose these controls before it can get into gasoline rationing.

This reimposition of controls may be tough. It has taken eight years to perfect this set of regulations. And still there are all kinds of problems; the line of people waiting to get exceptions or to make appeals from the process is very long still. I am not sure that DOE can reimpose that kind of a system at the same time it is trying to do gasoline rationing, to initiate this whole second currency. Maybe it can, but there is some question.

In the coupon rationing scheme, DOE would try to estimate how many gallons of gas would be available three months in the future. It would print and then distribute checks. Because coupons are transferable, they could not be sent in the mail. They are going to be worth money and, just as currency cannot be sent through the mail, neither can valuable coupons. Everyone who owns an automobile will receive a check in the mail that they will take to a bank. They will endorse the check over to the bank and either get dollars or coupons. The car owner then uses the coupons to buy gasoline. Let us say the interruption requires that the amount of gasoline be cut back so that you only get 10 coupons a week per car. If you use more than 10 gallons a week, then you must buy some coupons from someone who uses less. There are some obvious advantages to the flexibility achieved by allowing these coupons to be bought and sold. There are also problems with it. First, you are in effect creating a second currency with all of the accounting and banking procedures associated with the first currency. Coupons could and will likely be used to pay taxi fare, pay allowances, buy milk—legally, or otherwise. It is similar to money. A second plan of tradeable, transferable coupons allows a reasonably efficient distribution of available gasoline supplies, but that is all it does. That is, it only addresses gasoline. Presumably, although again this is not made explicit, gasoline would be made to take the brunt in a crude oil shortfall. Allocation would presumably be used to try to maintain undiminished the flow of other products—home heating oil, resid, jet fuel. Somehow, it seems to have been decided that gasoline is the discretionary fuel and that we will lay the burden of a shortfall on gasoline.

The third plan would try to substitute exactly for the coupon

rationing plan through a gasoline tax and rebate. Again, this plan
would only apply to gasoline and would load the entire shortfall on
gasoline. To put an excise tax on gasoline, you must again estimate,
first, how many gallons are going to be there, and second, what
market clearing price will be required to eliminate lines. Referring
back to the coupon gasoline rationing plan, clearly, the coupons will
not settle down in value. However, suppose that they settle at a value
of $2 a piece—that is, you can buy and sell them at that price. Then
the idea in the tax rebate plan would be to put a $2 excise tax on gas-
oline and mail cash, rather than coupons, to automobile owners.
This would have very similar kinds of effects, at least in theory, to
the coupon plan. In both cases, consumers would be facing the market price of $3.50. If, in the coupon plan, controlled gasoline sup-
plies go for $1.50 and the coupons are worth $2, then the market
clearing price really is $3.50 in the coupon rationing world. In the
tax rebate world, the market clearing price would be very similar.
The government would impose a $2 excise tax, and then would rebate
the $2 to the same people who would have received coupons. In
other words, if you had sent coupons to everyone who owns a car,
with a limit of three cars per household on the coupons, then you
would do the same thing on the tax rebate. Instead of sending out
10 coupons worth $2 each for each car, you would send $20 for each
car. The government would send you a check for $20, and you could
decide whether or not you wanted to go out and buy gasoline with it
or to use it to pay allowances or pay for taxis or whatever.

There are clearly some difficulties—how do you know what the
market clearing price is? That is a really hard question to answer,
but there are ways to get around it. Instead of putting on an excise
tax, it has been suggested to put a special windfall tax on margins of
refiners, jobbers, and retailers—everyone downstream—and there
would be a very similar effect without having to try to guess the
market clearing price.

The oil market has some interesting aspects for economists.
Some people have suggested that at the outset of an interruption,
when prices are going up very quickly, the elasticity of demand may
be positive. That is, as the price is going up, people somehow demand more. There is the panic aspect: people buy gas, or if you
are Hertz, you buy gasoline to put in your fleet's tanks, so that tomorrow, when gas is likely to be more expensive, you are better
off. There is also the investment notion: if you know the price is
going to go up during the interruption and you can buy low now, then
it makes sense, from just a rational point of view, to do that. These
two aspects of the effect of expectations may give us a positive elasticity in the very short term. The excise tax—if it could be put on in
a hurry—may be preferable to the special margin tax, in that people

are presented with an immediate jump in price, so they are not attempting to go out and beat the market and buy gasoline at its lower price now to avoid the higher prices tomorrow. They are presented with the "step function" instead of the "ramp function" in the price over time. Under the gasoline tax/rebate, controls on domestic crude oil are probably also needed. You may not need controls on downstream operations, that is, the margin controls and the allocation plan. But to keep the excise tax at a nonzero value, you have to again create a shortage, just as you do to keep the value of ration coupons at a nonzero value. Controls will do that.

The fourth plan can be called decontrol and general rebate. It would apply to more than just gasoline. For this option, you would not control the prices. This is the only one of the four plans where you would not have crude oil price control. You would allow all prices to rise with the rising world crude market. You would expect, under this case, that the government would all of a sudden have ballooning windfall profits tax revenues. If you did not put controls on domestic crude, then domestic crude would go up with world crude, presumably rising fairly quickly during the interruption. The windfall profits tax is an excise tax that takes a percentage of the difference between some base price and the market price. As that market price goes up, the percentage stays the same, leading to increased revenues. The government is going to have a lot of money on its hands if it does not control the crude oil prices. It can use these revenues to rebate in some manner. Note that all prices are going up: for example, truckers and farmers are paying higher prices for their diesel, therefore, it costs more to get food to market and food is priced higher. Therefore, stronger arguments can be made that the rebate should be general in nature—that is, distributed to everyone. There is no reason why automobile owners alone should get the rebate in this case. It becomes interesting to look at the effects of oil price rises on everyone, as opposed to just automobile owners.

Neither the rebate with the gasoline tax where the rebate would be sent to automobile owners nor general decontrol where the rebate would be sent to everyone is easy. You can try to conceive of getting a list of all Americans to mail checks to. Some people have suggested electronically transferring money to everyone's bank account. Still you need to have that list. You have got to know who has bank accounts. Everyone who does not now have a bank account has to get one (or two or three). There are other kinds of ways to rebate, but none of them are easy, and none of them are satisfactory in all respects. They all would take a long time to implement. There may be a way of sending money through existing programs to the people most directly affected, the people most likely to be put in life-threatening situations. The government may have an obligation to that group of people and there are existing programs that might be funded incrementally. These include low-income energy assistance, a general tax cut, a payroll tax cut, state tax cuts, or state grants to people in distress. Those kinds of rebates, while not going back to every individual (and, therefore, providing only rough justice at best) may be the only way to help people quickly.

Those are examples of the four kinds of plans DOE is considering. In either of the two plans involving the rebate, there are complications due to the Consumer Price Index (CPI). If you do not include the rebate in the measurement of CPI, then all indexed wage contracts and transfer programs will escalate with the higher gasoline prices. There may be a way of either including the rebate or excluding the tax when defining or calculating the CPI. Or maybe, as some people have suggested, you could index the rebate, that is, if you have a wage contract that is indexed to the cost of living or the CPI, then your rebate is somehow reduced. There are ways of dealing with the CPI, but some people see it as a big problem.

The coupon rationing plan and the gasoline tax/rebate hit only gasoline. Now this is something California in particular would be very interested in. It is not entirely clear why we should forego opportunities to conserve home heating oil, why we should forego opportunities to conserve residual fuel, why we should forego opportunities to conserve jet fuel; all of these conservation opportunities are foregone if we concentrate only on gasoline and allow its price alone to rise. You can calculate the dead weight losses associated with loading it on gasoline alone and they are high. Or you can also calculate what tax would be required to do the same kind of conservation, gasoline versus all products, and that is high. There may be strong reasons to take a look at all products rather than just gasoline.

The fairness criteria is "perceived fairness," and that may be distinct from what is actually fair. An example comes to mind: when the shortfall is concentrated on gasoline, the government is going to send coupons or rebates back to people who own cars. The idea is that people who own cars are the people who use the most gasoline. They are the ones who are going to be hurt the most during the interruption. There is another way of looking at it: gasoline is in some sense more valuable during this interruption and the price may stay high and it may not come back down. In this case you are talking about a commodity that is more valuable during the interruption and probably thereafter. It then is not clear why you want to try to make people "whole," who use more of that particular, more valuable commodity.

In addition, we are sending back three times the amount of money to people who own three cars than to those who own one, and infinity times the amount to those people who own zero cars. People with no car get zero rebate. On the other hand, of the four plans—the price controls and allocation with lines, coupon rationing, the tax/rebate, or the general rebate with decontrol—many people would say the gasoline rationing plan is the fairest. I am not sure that is, in fact, the case or that it would be perceived to be the case after some period of time of operating the gasoline rationing plan.

There are practical problems with each of these allocation plans. The largest practical problem associated with the first three plans is trying to reimpose the controls that have been on for the past 7 to 8 years. A big practical problem associated with the decontrol and general rebate is, of course, the rebate, unless you are willing to go with this very rough justice notion of sending back money through existing transfer programs.

The whole issue surfaced again in 1981 with consideration of renewal of the Emergency Petroleum Allocation Act. The question remains "What kinds of standby authorities need to be extended? What kind of standby emergency authorities does the Congress want to give to the president or DOE to have on the shelf in case of an emergency?" The initial response in Congress is "Well, we're just going to keep the standby authority to allocate and to price control, and then we are going to lay it on the states to take care of it." That is certainly one option; and maybe, no matter what we say or do, that is the authority that will be granted. But there may be a chance, as Congress is debating what other kinds of authorities to put in, to provide some flexibility: "Okay, you can control price and allocate, but you can also have the authority to put on an excise tax right away and send out a rebate."

16

LESSONS OF THE STRATEGIC PETROLEUM RESERVE

Carlyle Hystad

The initial plan for the Strategic Petroleum Reserve (SPR) program was to store crude oil. The goal was to store 500 million barrels of crude oil by the end of 1982 and 150 million barrels by the end of 1978. The Department of Energy (DOE) decided it ought to be stored in underground salt caverns, which were conveniently located on the Gulf Coast of Louisiana and Texas. There were a number of potential sites, some of which had existing capacity that had been developed both through solution mining and through conventional mining. Five sites that had existing capacity of approximately 250 million barrels were selected. On two of those sites there was room in the salt dome to leach (or solution mine) additional caverns to get up to about 500 million barrels on these first five sites. In 1980, those five sites were selected, purchased, and developed, and are now ready to hold 250 million barrels. However, early in 1980, there were just a little over 91 million barrels in those sites.

These sites all have the capability in time of interruption to move oil out to both tanker docks and existing crude oil pipeline systems that now take a large share of the imported crude oil that comes into the Gulf Coast and up into the interior of the country. The SPR has direct connections to the three major crude oil pipelines that carry oil into the interior, plus there are connections by pipeline to a number of the local refineries in the Gulf Coast area—which, of course, is the major refining complex in the country. It also has direct connections to approximately 15 tanker docks that could be used to load SPR oil and move it to any other site that now receives imported oil by water. When these sites are fully developed up to 500 million barrels, SPR would have a capability to draw down approximately 3 million to 3.5 million barrels per day (b/d) from those five sites.

The plan was modified with the advent of the Carter administration. The schedule was accelerated to try to have 250 million barrels by the end of 1978, 500 million barrels by the end of 1980, and 1 billion barrels as soon as possible thereafter. It would have been necessary to select additional sites and the plan was to proceed essentially on the same basis and select three or four additional sites that would be tied into the same three basic distribution systems.

The plan was wrong in several ways. First, there was never any realistic plan for achieving the accelerated development schedule. The schedule was arbitrarily set and there simply was no way of doing it. There was no technical ability to achieve that objective, but a lot of attention was diverted away from developing these first five sites to try to find some way to do the impossible. As a result, a slowdown of the program occurred.

Probably the biggest adverse impact was the creation of DOE. It was created at the worst possible time in that the program was just well under way. The sites had been acquired.

We were in the midst of awarding contract and starting the construction work. With the creation of DOE on October 1, 1977, about 25 percent of the key personnel of the SPR program were pulled out to deal with the more important DOE problems, even though the secretary and the president had said that the SPR program was the most important in the government. There also was a freeze on hiring in DOE that lasted for several months. In addition, a whole new set of hiring procedures was developed, so even when the freeze was lifted, it was very difficult to hire the new people needed to move the program. As a result, all procurements came to a halt and a whole new procurement system had to be installed. Under the Federal Energy Administration (FEA) program, procurements had been moving through in about six weeks, while the schedule in DOE was for about nine months from start to finish. Essentially, no procurements were awarded for the first several months in the new department. Most significantly, priority was given to cost rather than time schedule, which some believed was a classic example of false economy. The government was concerned about saving 10¢ a barrel in construction costs here and 10¢ there. What happened, of course, is that now the cost increased to $20 more for a barrel of oil because of the Iranian interruption.

There were several other things that had some impact on the program. There were problems with environmental permit delays with the state of Louisiana and with the Environmental Protection Agency (EPA). However, they were not a major problem and have been used as a scapegoat to explain the delays when they really were not the primary reasons for delay. The same is true of real estate acquisition. There were some problems in getting condemnation of real estate, particularly for pipelines, that may have set the program back a few months, but, again, they were not a key factor in programming.

There were attempted corrections. First, in order to deal with the procurement problem, it was decided to set up an office in New Orleans to take responsibility for all the procurement, in order to get it out of the headquarters where everything was at a complete standstill. Unfortunately, setting up an office in New Orleans took about six months. In addition, there were a lot of people who did not want to go to New Orleans. In the long term, the decision to move was the only solution, and indeed the office is now functioning quite effectively. A whole new set of cost control procedures were put into effect that resulted in many long, agonizing decisions about whether to make some changes that may increase costs. These led to further delays.

Probably the biggest mistake was solicitation of turnkey proposals. Someone had the bright idea that the way to solve the problem was to go to industry for fixed price proposals to develop a site and then to turn it over to the government to be filled. There apparently was an assumption that industry would be willing to take the risk and do this at a very low cost. Others believed that was a very naive assumption, and after spending about a year in soliciting these proposals and reviewing them, the government came to the conclusion that, indeed, it simply could not afford to pay the price that the private proposers wanted for developing these turnkey sites. The prices were about five times the government cost to develop the sites. So, essentially, a year was wasted and no work was done on selecting any new sites.

So what are the continuing problems and where does the program stand today (1980)? Before this turnkey effort was completed, the Iranian interruption occurred and then there was the new problem of oil availability. That problem has not been resolved. Because of this there has been no action taken even to think about going beyond the first 500 million barrels. Everything essentially is at a standstill in terms of selecting sites beyond the first five developed sites.

The program remains inflexible and there is an ineffective procurement process. This has become more important now for the oil procurement system rather than for site development. The open competitive solicitation for oil simply does not work in the kind of market we have been in since the Iranian revolution. In addition, we cannot expect it to work effectively in the future except in very temporary situations. Nevertheless, there has been no alternative developed, even though there has to be the alternative of letting the Strategic Petroleum Reserve (SPR) purchase oil in an essentially noncompetitive fashion and probably even in a classified or secret fashion.

Almost since the beginning of the program, there has been opposition from the Office of Management and Budget (OMB) in the White House. This opposition has been particularly strong to expanding it beyond 500 million barrels. OMB continues to oppose this (1980), so a great deal of staff time in the SPR office as well as other parts of DOE is spent in continued cost/benefit studies for OMB.

Finally, there is continued disagreement on the appropriate role for the SPR. When will it be used? How shall it be used? What are we trying to achieve with it? There are some who feel it ought to be drawn down immediately, even in a small interruption, while others feel it should be reserved only for the most severe interruptions. In 1979, in the height of the gasoline lines around the country, there was a large meeting of most of the agencies in the White House. The discussion of whether or not the SPR should be used was one of the few times that just about everyone agreed that it should not be used for that type of an interruption. There was a very broad consensus that the SPR should be built up and held for potentially more severe emergencies.

What are the prospects for the future? I believe that we will never have a large strategic petroleum reserve because of problems in purchasing oil, problems in getting new sites, difficulties in making decisions, getting funding, and so forth. It will get delayed so long that the problem will go away or alternatives will be found. It will be several years before we have an SPR that even will be of real use for any interruption. An estimate is that if we are lucky, by the end of 1982 or early 1983, we will have the existing capacity filled, that is, up to just under 250 million barrels. The present best estimate is that we will not reach 500 million barrels until about 1987 or later, and that may very well slip another year or two because the second 250 million barrels needs to be developed. It requires solution mining of the salt caverns and that is a slow process. Even if they are able to buy the oil to fill the second 250, as the capacity is ready, it will not be completed until about 1987 at the earliest.

A small SPR (in the 300 million to 400 million barrel range) will not be used in a small interruption such as 1979. The decision made in 1979 will continue to be the kind of decision made in the future for these interruptions. What we really ought to do is recognize that fact and earmark a strategic reserve, a national security reserve, that is set aside for people to use if ever they decide things are that serious, and develop a second reserve to be used to deal with smaller interruptions. But the development of that second reserve does not seem feasible.

The five sites that have already been selected for the SPR should be this national security reserve, this doomsday reserve, and we should look to private sector storage to be the useable, the useful, the tactical reserve to deal with the smaller interruptions. Authority now exists to require refiners and importers to store up to 3 percent of their annual throughput, which would provide about 180 b/d of private sector storage. It was decided not to do that for a number of reasons. Probably one of the most important reasons was the great difficulty in enforcing and verifying the requirement and determining whether people have indeed increased their storage and whether they are maintaining that oil. The question of how to assure that the oil is being used during an interruption still remains. In addition, it was felt that it would be higher in cost than government storage in underground caverns, and there were also some legal concerns about the way the act was written.

If there is going to be a private sector storage program, it has to be a program that solves the problem of the verification of buildup and drawdown. We must be able to assure that we can draw down, otherwise everyone's time and money are being wasted. It would appear that there are two possible options for a private sector storage system that could be effective. One would be patterned after the German system, which is in essence a private storage corporation that has the authority to build its own storage, essentially tax the oil companies, the refiners, importers, or whatever other group of industries you wish to tax, to pay the annual carrying cost of developing and maintaining the storage. The storage corporation would be able to operate free of the federal personnel and procurement requirements, so as to get away from all the problems that the SPR has in those areas. It probably would be necessary to have the board of directors of that storage corporation in a position to make the decision on drawdown independent of any government. The government may want to be represented and ought to be represented on it, but not in the majority position. Rather there should be a board of directors with representation from state and federal governments, the oil industry, and users.

Another option that is worth considering is to have a requirement placed on oil companies or major users that they have the capability to draw down petroleum stocks by a certain amount during an interruption. The focus would not be on trying to determine how much stock they have in mind. Rather the requirement would be placed on the drawdown, not on the buildup. There is some legislation where the oil companies or major users would be required to have in place a capability to draw down, say, so many days of their preinterruption throughput. The failure to draw down, if they are directed to do so, would result in a substantial financial penalty for each barrel that they are short or are unable to draw down from their stocks. The advantage of this particular approach is that it would require no effort to verify stock in advance of an interruption. It also would not require the federal government to establish a whole bureaucracy to determine and designate what particular type of storage or oil ought to be stored and would not require the government to tell the oil companies specifically how to meet their drawdown requirements. That decision would be up to them. Their only requirement would be to reduce their stocks in aggregate by a specified amount that could be verified by comparing the input and output of the firm rather than trying to verify changes in stock levels.

17

PRIVATE SECTOR PETROLEUM STOCKPILING: THE FEDERAL AND STATE CASES

Susan L. Missner

Holding inventory is a time-honored response to the problem of continued reliance on an unstable source of petroleum supply.[1] During the oil embargo of 1973-74, the United States, recognizing the insecurity of its primary source of oil, the Persian Gulf, passed the Energy Policy and Conservation Act (EPCA). The legislation authorized the creation of a crude oil stockpile providing storage of at least 150 million barrels by the end of 1978.[2] President Carter embraced the Strategic Petroleum Reserve (SPR) as a central plank of his energy policy and accelerated the SPR fill goal to 1 billion barrels by 1985. Then, citing budgetary reasons, he deflated the fill target to 500 million barrels by 1981 and 750 million barrels by 1984.

In 1980 the SPR had less than 92 million barrels of oil, representing about 12 days of imports. No oil purchases have been made for the reserve since March of 1979. At the minimum 100,000 barrels per day (b/d) fill rate ordered by Congress in June 1980, it would take until the end of the century to create a 1 billion barrel stockpile. The filling of the SPR is at least being delayed and its future is uncertain.

Recent analyses at Stanford University[3] and the U.S. Department of Energy (DOE)[4] show the high importance of having a petroleum stockpile as a buffer against the adverse economic impacts of a major oil supply cutoff. With the large GNP losses associated with a disruption of more than minor depth and duration and the federal SPR at a trivially low level, attention has been focused on the alternative of private sector stockpiling options for the country. Along with the creation of the SPR, EPCA appears to provide adequate basis for a substantial program of private sector stockpiling. The legislation gives DOE the authority to establish an Industrial Petroleum Reserve (IPR) as part of the Strategic Petroleum Reserve. Under this authority,

DOE may require each oil importer and each refiner to "1) acquire and 2) store and maintain in readily available inventories" petroleum products in an amount no greater than 3 percent of the last calendar year's imports or refinery throughput, as the case may be. In its "master" SPR Plan presented to Congress in 1975, DOE stated that industrial storage of petroleum products "would not contribute to more efficient development of the SPR nor result in any significant cost savings to the nation at any stage of implementation of the Reserve."[5] If the "3 percent rule" was implemented based on imports, the statute would permit the storage of about 90 million barrels in 1980. If based on refinery throughput, it would permit stockpiling of about 150 million barrels.

THE U.S. STRATEGIC PETROLEUM RESERVE

The conventional benefits of an oil stockpiling policy have been widely cited. A stockpile can prevent huge losses to the GNP that would otherwise occur in the event of a supply disruption. It also can reduce panic hoarding during disruptions and can act as a deterrent to discourage producing countries from intentionally disrupting supplies by reducing the effectiveness of such an action as a political or economic weapon. In addition, it can buy time in a crisis, thereby preventing a premature escalation to military confrontation, and it can ensure an emergency supply of petroleum to the military in the event that an armed conflict does occur. However compelling these benefits may seem, the substantial gap between stockpiling policy and practice in the United States is testimony to the lack of political potency of the program supporters once a crisis is over.

The SPR has been subject to the implementation problems and political haggling that often plague public sector projects. The Plan authorizes development of storage capacity at five sites along the Louisiana and Texas Gulf Coast with accessibility to both interstate and crude oil distribution pipelines and port facilities. Although salt dome storage is conceptually simple, development of the five sites has proven to be an engineering and management effort of major proportions.[6] The schedule has not been met because of delays in land acquisition (there are thousands of landowners involved), contract awards (most industry participants in SPR equipment and construction were unfamiliar with government procurement processes, resulting in a limited number of bids), staffing, and capacity utilization.[7]

Though the SPR had little, if any, political opposition nationally, it ran into political problems of an international nature. In the spring of 1979, U.S. government sources began leaking to the press that Saudi Arabia was privately threatening to cut its oil production by

1 million b/d if purchases for the reserve were resumed. The Saudis maintained that their reserves could be considered the West's strategic stockpile and that a U.S. storage program was therefore unnecessary. They claimed that U.S. stockpiling purchases add to world oil demand and oil price pressures, undercutting the Saudi efforts to moderate and unify OPEC's pricing policy. DOE was thus placed in the uncomfortable position of explaining away SPR purchases. It did not want to pressure world demand and prices nor irritate the Saudis (whose 1 million b/d production increase was helping to soften the shortfall caused by the plummet in Iranian production), and reserve purchases were stopped or "delayed" as government sources claimed, until the world oil situation was softer and more conducive to storage purchases. Further, the State Department argued it was bad diplomacy to resume SPR purchases while it was urging allies to boycott Iranian oil, thereby forcing them to seek supplies elsewhere.

INDUSTRY STOCKPILING

On March 9, 1980, the New York Times reported that the Western industrial economies were stockpiling oil supplies at record rates. The International Energy Agency (IEA) was reported as estimating in 1980 that U.S. stocks would be up 11 percent over the previous year. Noting that this stockpiling drive was primarily a private sector phenomenon, having little to do with government stockpiling policies, the Times explained:

> . . . Normally, with steady output, a mild winter and sagging domestic economies, stockpilers tend to draw down on oil stocks already accumulated rather than buy more oil. But with the supply shocks of last year and the unsure state of future pricing and production levels by oil producers, the oil and industrial companies are continuing to increase reserves rather than liquidate their stocks.[8]

In 1980, there was a reported stock of crude oil and products in the United States of 1,076 million barrels (990 million barrels are working inventory)[9] along with an average of approximately 100 million barrels of oil en route from the Persian Gulf to the United States. In 1979, the reported stock was about 800 million barrels. In effect, private decision makers acted so as to increase the nation's oil stocks from the equivalent of about 100 days of imports to about 153 days (or, alternatively, from about 53 days of total consumption to about 71 days). At 1980 prices, this inventory represented an investment of about $32 billion and at prices that might prevail in a severe crisis, it could have a value of upward of $200 billion. This level of stock reflects, as the Times stated, expectations about the future that have recently been prevailing. Expectations change and the near future may see these stocks increase further or, on the other hand, be depleted to the minimum level required by the logistics of supplying oil to consumers.

The companies finance these reserves themselves. Energy-intensive industries, for example, may stockpile to prevent a production shutdown. They have an incentive to invest in storage, pay oil companies to store for them, or purchase future contracts for energy. Consumers of energy-intensive products pay the stockpiling costs that are passed through to them. If such costs are too high, this will be reflected in the market and be a signal that the cost of ensuring an uninterrupted supply is worth less than the benefits to the stockpiling firm. Because some of the benefits to the nation of having an oil stockpile cannot be captured by individual firms (e.g., the possible deterrent effect on supply interruption or the prevention of macroeconomic disturbances from oil supply interruptions) and because the applicable discount rate for society is arguably lower than for firms, private stockpiling of oil will be systematically less than is optimum for society. The question now is whether government action working through the private sector can achieve what neither the government nor the private sector are able to do working separately.

A REQUIREMENT FOR HOLDING
A MINIMUM RESERVE

If the government imposed a legal requirement that companies importing oil must hold a minimum level of stocks available for use in the case of an emergency, the country would have greater protection against a major supply interruption. This industry-holding option is described in Energy: The Next Twenty Years[10] and amplified by Henry Rowen and John Weyant.[11] The minimum level of imported oil and products would be held in stock by importers who could contract with anyone in the United States to store it.[12] The holding costs would be a cost of doing business and the economic impact would be equivalent to a tariff on imported oil. This cost would be an appropriate way of internalizing the insecurity cost of dependence on unreliable Persian Gulf Oil. Private stockpilers would have an incentive under this plan to find the lowest cost means of holding the required amount. The required minimum stock level could be raised each year until the required overall level, taking account also of the amount in the federal reserve, is reached. The target stock level for the two types of stocks together might be the administration's proposed level of 1 billion barrels.

The marginal source of imported oil in the Persian Gulf and dependence on this source bears with it a substantial and costly insecurity premium. [13] Therefore, the users of this oil should pay an "insecurity" premium. If all importers are required to bear these costs, competitive forces should leave none at a disadvantage. (This analysis assumes an end to the federal entitlements program that equalizes the cost of crude to refiners independently of the source of their oil.) The domestic price of oil will rise to the level determined by the price of imports, which will include the cost of holding stockpiled oil. Now that the windfall profits tax is in place, the government will receive most of the increased revenues from the higher domestic oil price. [14]

For example, if marker crude on the world market is priced at $30 per barrel and the capital cost per barrel stored is $10, the annual cost to a firm of holding oil beyond the level it desires for business reasons would be around $7 per barrel at an interest rate of 15 percent (assuming the oil is used as collateral). If an interim target level of 200 million barrels were adopted for this privacy reserve above expected minimum stock levels of importers, the capital investment would be $8 billion and the annual carrying cost $1.4 billion (neglecting costs of managing the stockpile and amortization of the storage site). [15] This would be equivalent to about $.47 for each barrel of oil imported now (1980). It would require a fill rate of about 275,000 barrels per day during the 1981-82 period to attain this level.

Although the principle of internalizing insecurity costs would be violated, some of the costs incurred by compulsory industry-owned stocks could be subsidized through government aid in the form of government loan guarantees or other subsidy methods. The argument for providing a subsidy presumably is that dependence on insecure, imported oil is a national problem and the cost should be borne by taxpayers at large (and also that a lower interest rate is more appropriate for oil stockpiling than the private market one). For example, a subsidy could be applied to the unused SPR capacity of DOE's storage facilities. The 91.7 million barrels of oil already stored represent 37 percent of the reserve's currently available storage capacity. To get oil accumulated there, DOE could offer to store oil belonging to the oil companies in these facilities at DOE's incremental storage cost (as long as it does not interrupt the filling of the SPR). The companies would bear the larger burden of costs—carrying charges for the fill. Perhaps future oil increases will offset some or all of the costs anyway. Furthermore, temporary storage by oil companies in the SPR facilities offers them the opportunity to sell oil to the SPR at a later date.

The advantage for the government under this plan would be the experience in filling and drawing down stock during normal times (since the SPR would not just sit until a crisis struck), allowing the identification of operating problems and potential solutions before an emergency. The storage plan also allows the industry's emergency reserves to be segregated from its commercial inventories. In most foreign countries, the industry's emergency reserves are held with the commercial inventories. These amalgamated reserves have many drawbacks.[16] There is a great deal of uncertainty surrounding the levels of emergency reserves in the total stockpile and whether they would be available in the event of a crisis. Edward N. Kraples states:

> . . . in countries where storage obligations have been in force for some time, the oil industry may have become used to operating the distribution system with a 90 day minimum. Over time, the system becomes less efficient; that is, it needs a higher level of working stocks because there is no penalty for inefficiency. Also, seasonal preparations by the oil industry influence national emergency reserve levels. Companies may even come to rely upon larger and larger seasonal stockpiling because the minimum stockpile requirements make it more economical to store a product than to increase production in the peak season.[17]

The problem, simply stated by Kraples, is that over time, a company will "absorb" the reserve, using it for commercial purposes rather than as an emergency reserve. By having the companies store their emergency reserves in the SPR facilities, this absorption trap is avoided. There is also an inherent advantage in having the strategic and industrial emergency reserves centralized under one management. Amalgamating the emergency and commercial reserves implies each company managing its own emergency reserves, a logistical nightmare from the perspective of a strategic reserve.

Obviously, an equitable industrial reserve obligation needs to be arranged. Imposition of the 3 percent rule, for example, would result in enormous inequities (and subsequent lawsuits) since all companies do not have similar financial structures. One possible solution is to vary the reserve obligations depending upon the company's size and capacity. A sliding scale could be established in order to discourage deliberate underproduction so as to escape a higher reserve obligation.[18]

CENTRAL STORAGE CORPORATION

Perhaps the most equitable method of establishing reserve obligations for industry is to establish a central storage corporation, thereby removing the emergency reserve obligation from the industry's books. The corporation would buy that amount of crude and products over the level the industry needs as its working inventory. It would borrow the money to buy the oil, using the stocks as collateral and purchase or lease storage facilities.[19] It would also administer the stocks. The corporation could pay the loan interest with compulsory uniform storage fees collected from oil-importing companies and based on the companies' sales volume, effectively passing the stockpiling costs to the consumer. Such a system could be adopted by levying a fee on imported oil, or from the revenues generated by the windfall profits tax.

THE STATE SPECIFIC CASE

Moving from the federal scenario to the state scenario, two issues need to be addressed: whether such stockpiling options as those outlined above exist on the state level and, if they do, whether it is even useful to implement them. Clearly, the scale for measuring such options is different when addressing the country as a whole versus an individual state; what may be a politically and economically sound policy for the nation may be neither good nor workable for the state. This is particularly true when the key players are multinational companies dealing in a product as fungible as crude oil and operating in a complex and volatile world oil market.

CALIFORNIA CRUDE OIL SKETCH

California is the fourth largest oil and gas-producing state in the nation. Sixty-five percent of the state's energy requirements are met by oil. With a daily production rate of 1 million barrels, in-state production satisfies 40 percent of this demand. Alaska contributes another 40 percent and a small percentage comes from the U.S. Gulf Coast states. As recently as three years ago, California was importing more than 40 percent of its crude oil requirements from foreign sources, but with the onset of delivery from the Alaskan fields, the state reduced its foreign imports to 20 percent. These are mostly from Indonesia, which has proven to be a stable source of supply for California over the last ten years. Further, Indonesia has made it clear it is eager to continue its relationship with the state of California, at least at the current level of need, if not at an increased level of exports to the state (thereby reducing its exports to Japan, its other major crude customer and a partner with whom the Indonesians are not entirely pleased).

Further, California has what are probably the world's largest heavy oil deposits. Estimated at 40 billion to 50 billion barrels, in today's economy, with gradual decontrol of domestic oil prices and OPEC prices at around $30 per barrel, it is believed that 6 billion to 12 billion barrels of heavy oil are recoverable today. Since most of the 35 refineries in California were designed to handle the large amounts of light foreign oils upon which the state has built its economy, California has mostly exported its heavier, high-sulfur crude while importing the lighter, low-sulfur crudes.[20]

It is generally asserted and widely believed that the probability of an oil supply disruption occurring during the 1980s is very high. It is also widely believed that unlike the targeted embargo of 1973-74, in which all OPEC members took concerted action, the most likely scenario generating another supply interruption will be internal chaos within the Persian Gulf region or the projection of power into that region by the Soviet Union. Either of these possibilities leaves Indonesia in a relatively stable position and so, tangentially, California's foreign crude oil supply.

A Private Petroleum Stockpile in California?

As stated earlier, holding inventory is a traditional response to the problem of continued reliance on an unstable source of supply. The United States indeed has a supply insecurity problem; about 50 percent of the country's crude oil requirements are met by foreign suppliers, 25 percent of that oil coming from the Persian Gulf. Given California's relatively secure oil supply network, is stockpiling oil worth the political and economic cost to the state?

Unlike the federal government, there is not, at present, any law on the books authorizing stockpiling in California. There would have to be an extraordinary demonstration of political will in order to compose and pass into law a stockpiling mandate.[21] How might such a law read? Adopting the reasoning used in the federal case, the state could impose a reserve obligation on all companies and refineries operating within the state's boundaries. For example, if the 3 percent EPCA rule was applied and based on refinery throughput in the state, 16 million barrels could be stored; if applied to the state's crude imports, 3 million barrels could be stored; and if based on the state's product imports, 472,000 barrels could be stored.[22] By imposing such reserve obligations, the state would be faced with the same equity problems faced by the federal government, i.e., imposition of an across-the-board reserve rule would result in enormous inequalities since companies have different financial structures. The reserve obligation could be varied on a sliding scale, depending upon the company's size and capacity, in order to avoid deliberate under-production and therefore a higher reserve obligation.

The most equitable method of establishing reserve obligations for industry is to establish a state central storage corporation, the formation of which would result in erasing the reserve obligation from the industry's books. The state corporation would buy that amount of crude and products obligated by the companies (according to their assigned reserve obligation), borrowing the money to buy the oil (using the stocks themselves as collateral), and purchase or lease storage facilities. The corporation would also administer the stocks. Presumably, the corporation could pay the loan interest with uniform storage fees collected from the member oil companies based on the companies' sales volume within the state, effectively passing the state stockpiling costs to the consumer.

There are some problems with the state adopting such an approach in an effort to protect its economy from the adverse impacts of an oil supply disruption. The fact that no legislative precedent exists within the state could make it difficult to pass such a law. Second, the state does not have any storage facilities at its disposal (unlike the federal government), and would thus have to build caverns which which to store the emergency reserve, a terribly costly and time-consuming project, even taking into consideration lessons learned from the federal government's foray into this business on the coasts of Texas and Louisiana.[23] A seemingly simple solution is to allow the companies to amalgamate their state emergency reserve obligations with their working inventories. The trade-off for such a solution, however, would present difficult logistics and management problems for the state. As Kraples points out, over time the companies "absorb" these reserves, using them for commercial purposes rather than as emergency reserves. At the national level, the management and administration of such a decentralized storage system would be so costly as to the point of impracticality. At the very least, it would involve monitors at all the storage (company) sites.[24]

A more manageable system might allow the state to impose the legal requirement that companies importing oil from the Persian Gulf must hold a minimum of stocks for emergency use by the state. Since oil from the Persian Gulf bears with it an "insecurity" premium, users of this oil should pay that premium. If all importers bear these costs, competitive forces would leave none at a disadvantage. This plan would, theoretically, give the state greater protection against a supply crisis. The minimum level of imported oil and products would be held in stock by importers who could contract with anyone in the state to store it.[25] The holding costs would be the costs of doing business and the economic impact would be similar to that of a tariff on imported oil into California. This would be an appropriate way of internalizing the insecurity cost of importing unreliable supplies. The stockpilers would have every incentive under such a plan to find the lowest cost means of holding the required amount.

The state could offer the companies incentives and subsidies in order to alleviate some of the extra cost. As in the federal case, the state could offer such assistance under the argument that Persian Gulf imports into the state represent a statewide problem and thus the cost for protection should be borne by the taxpayers of California. The state could offer loan guarantees and tax write-offs as well as exempt presently held stocks from any state reserve obligation.

Does such a plan make sense for California? In fact, the state has reduced its imports from the Persian Gulf region (the insecure suppliers) to a negligible amount, thanks to crude coming in from Alaska.[26] Persian Gulf imports into California are mostly represented by products from the U.S. Gulf Coast region that originated in the Persian Gulf. As calculated earlier, a 3 percent reserve obligation based on California's product imports would result in less than 500,000 barrels, indicating the small, targeted amount. Due to the nature of the world oil market, it would be difficult to track which products originated in the Persian Gulf and hardly worth the effort given the small amount California imports from that area. Its remaining foreign supply source, Indonesia, is not an insecure source as defined here. Imposing reserve obligations on companies importing Indonesian crude (keeping in mind the tariff-type effect) would only antagonize a heretofore friendly source of supply.

CONCLUSION

California is blessed with one of the healthiest and most diversified economies of any state in the nation. It has also proven itself capable of taking the lead on issues of political, economic, and social concern to the country as a whole. But because of the complex nature of the energy problems the country faces, as well as the global intricacy of the oil market, it appears that the energy policy arena is one in which the states, acting on their own, have very few options available to them. Two options that are available to California to strengthen its security of energy supplies would be purchasing more oil from Indonesia and less from Alaska, thereby freeing up Alaskan supplies to be sent eastward. The East has a higher import dependence, particularly on insecure sources of foreign supplies.

As regards more general policy initiatives like a state stockpiling program, there is little room in which California can maneuver. Such initiatives are optimal at the national level, not the state level, due to the nature of the problem and the industry. The utility industry in California provides a good juxtaposition to the point. The utilities store oil to first, assure customer supply; second, as a hedge against the wide swings in hydro availability; and third, for contingency. Each utility has a certain number of days of fuel storage authorized in its rate base by the California Public Utilities Commission (CPUC). The utilities in the state are regulated by the CPUC and are therefore subject to its rules. This is not the case for the oil companies operating within California. Rather, they are privately owned and operated and not subject to the rules and regulations of state agencies. They are regulated at the federal level, however, which is where the authorization for stockpiling should originate. It is not workable for only one state to mandate private stockpiling within the petroleum industry—it is good for all states to do so. They can assist by lobbying the federal government to mandate such a reserve.

NOTES

1. The first few pages of this chapter closely follow the line of reasoning used in "Private Petroleum Stockpiling Options for the U.S.," July 1980, prepared for the U.S. Department of Energy.
2. Energy Policy and Conservation Act, PL 94-163, December 1975.
3. See, for example, John Weyant and Henry Rowen, "The Optimal Strategic Petroleum Reserve Size for the U.S.?," Stanford University International Energy Program, October 1979; "Reducing Dependence on Persian Gulf Oil: Needs and Opportunities," Stanford University, March 1980; and "The Problem of Security of Supply of Persian Gulf Oil: Analysis and a Proposed Strategy," April 11, 1980.
4. "An Analysis of Acquisition and Drawdown Strategies for the Strategic Petroleum Reserve," U.S. Department of Energy, Office of Oil Policy, December 1979.
5. Energy Policy and Conservation Act, PL 94-163, Section 6236.
6. Problems in SPR implementation are detailed in the Annual Strategic Petroleum Reserve Report, U.S. Department of Energy, Strategic Petroleum Reserve Office, February 1979, pp. 22-24.
7. In order to permit continuation of sale production at a Morton Salt Company mining operation and avoid unemployment of 300 workers, the government had to absorb an 11-month loss in the SPR's planned construction schedule.
8. John M. Geddes, "Western World Found Stockpiling Oil," New York Times, March 9, 1980.
9. Oil and Gas Journal, March 31, 1980.

10. Resources for the Future, Energy: The Next Twenty Years (Cambridge, Mass.: Ballinger, 1979), pp. 38-39.

11. John Weyant and Henry Rowen, "Reducing Our Dependence on Persian Gulf Oil: Needs and Opportunities," Stanford University, March 1980.

12. This level would be composed of a representative mix of oil and products so as to avoid storage of the cheapest crude type.

13. Although any component of the world's oil network could be regarded as insecure (e.g., the Alaskan pipeline, Caribbean refineries, and offshore oil platforms are all subject to sabotage), using the Persian Gulf effectively bounds the insecurity argument since it is the marginal source of imported oil.

14. An alternative is to have all oil companies stockpile a percentage of their imports or throughput, as the case may be. This alternative, and its management and equity problems, are discussed later in this chapter.

15. If firms contracted for use of the government's already existing SPR sites, some of this initial capital cost might be avoided. On the other hand, this estimate of costs assumes no increase in the world price from this stockpiling. If a tight oil market exists during the filling period, the effective cost of the stockpiled oil, in private hands or in the SPR, could be much higher.

16. For an in-depth discussion of the amalgamated versus segregated stocks issue, see Edward N. Kraples, "Proposed Guidelines for an Industrial Petroleum Reserve Policy," working paper for the U.S. Department of Energy, May 1979. Also see E. N. Kraples, EL-78-X-01-4848, U.S. Department of Energy, September 1978.

17. Kraples, "Proposed Guidelines," p. 7.

18. Kraples, "Proposed Guidelines." He points out the inequities by examining the interesting, though unfair, possibility of exempting small firms from the reserve obligations. Of 150 refining companies in the United States, 28 control 85 percent of the national total refining capacity. If one imposes the 3 percent rule on all 150, a 200 million barrel reserve obligation results; if imposed only on the top 28 companies, a 175 million barrel obligation results.

19. Though the banks would accept the stocks and facilities as collateral, they are unlikely to provide a dollar of credit for each dollar of current value represented by that oil at the going market rate. The federal government would need to guarantee this collateral "gap."

20. California Energy Commission, Biennial Report (1980), chap. 2. In the near term, the heavy oil deposits could not directly enhance California's supplies during an energy emergency. California could continue exporting the heavy oil, presumably at higher prices, thereby bringing that much more revenue into the state for light oil purchases.

21. If for no other reason the oil companies would presumably vehemently oppose any such plan.

22. Based on California Energy Commission data for 1979.

23. Before 1978, there was some storage space available; since the Iranian revolution, however, oil stocks are at an all time high. The major oil companies claim little, if any, storage space exists at this time (1980).

24. Though at the state level, spot checking may be slightly easier.

25. Of course, the stock would be composed of a representative mix of products so as to avoid storage of the cheapest crude type.

26. As mentioned earlier, the Alaskan supply system is prone to sabotage or earthquake damage. But damage to the pipeline itself could be repaired relatively quickly.

18

FINANCING OPTIONS FOR THE STRATEGIC PETROLEUM RESERVE BY THE UNITED STATES CONGRESSIONAL BUDGET OFFICE

U.S. vulnerability to disruptions in imported oil supplies led the Ninety-fifth Congress to authorize the creation of a Strategic Petroleum Reserve (SPR) in the Energy Policy and Conservation Act (EPCA) of 1975. The SPR could be highly effective in mitigating some of the adverse economic effects associated with supply interruptions. One projection, for example, indicated that in the absence of a reserve, a yearlong shortfall of 2 million barrels per day in 1984 would reduce projected GNP by approximately $146 billion (3.6 percent) and increase the unemployment rate by 1.1 percentage points and the inflation rate by 7 percentage points. Drawing down a 750 million barrel reserve could avert virtually the entire impact of such a disruption.

These benefits notwithstanding, the SPR program has experienced numerous difficulties and delays. Of the 1 billion barrels of storage capacity authorized by the Congress, only 250 million have been completed, with 150 million more slated for completion by 1985. Furthermore, the reserve now contains only 121 million barrels of oil. In response to the tight world oil market caused by the Iranian revolution, the Department of Energy (DOE) suspended oil purchases in February 1979, only to resume them at the direction of the Energy Security Act passed by the Congress in 1980.

FINANCING THE RESERVE

Administration plans for 1981 call for a fill rate of 200,000 barrels per day for the remainder of fiscal year 1981, and 230,000 barrels per day in fiscal year 1982. Together with the 121 million barrels of oil now in the reserve, this would create a reserve of

250 million barrels by the end of 1982. Maintenance of the latest administration plan for filling the reserve, which averages about 195,000 barrels per day over the next seven years, would create a 750 million barrel reserve by the end of 1989.

Filling the reserve under this schedule would require total additional budget authority for oil of $36.7 billion in fiscal years 1981/89. A supplemental appropriation is necessary because of the cessation of the entitlements benefits the reserve received while oil price controls were still in effect. Annual budget requirements for the SPR fluctuate with the planned rate of fill, rising from $4.4 billion in fiscal year 1981 to a peak of $7.4 billion in fiscal year 1987. The total cost of a 750 million barrel reserve, including the funds appropriated to date, is estimated to be $44.8 billion.

Together, these significant budgetary impacts and the compelling benefits of the SPR have recently focused attention on methods to attract private funds into the SPR program. Four of these financing options are analyzed here:

- Option one is the public capitalization of the SPR or SPR "certificates." This option would allow private investors to speculate in oil by purchasing title to the value of a specified quantity of oil in the reserve. The return on the investment would depend solely on increases in the value of oil, and could be realized by trading the certificates on a secondary market, redeeming them upon reserve depletion, or holding them to maturity.

- Option two is debt financing of the SPR. This alternative could be achieved by the federal government through the issue of a new series of bonds, the sole purpose of which would be to raise money for the SPR program. The new debt instrument could bear either a market rate of interest or a yield related to the rate of oil price appreciation.

- Option three is the development of an Industrial Petroleum Reserve (IPR). This type of financing would shift the focus, and much of the burden, of the oil stockpiling program to the private sector. Several policies could result in an IPR. Using the authority provided by the Energy Policy and Conservation Act (EPCA), the president could require all importers and refiners to store up to 3 percent of their consumption in a separate emergency inventory. Alternatively, the government could provide financial incentives, tax credits, or direct subsidies, for example, to firms increasing their inventories. Another plan would use the EPCA authority, but would allow those firms required to store oil only to show evidence that someone is storing the oil for them. By using this last plan, some speculative capital might be attracted from outside the industry.

- And option four is mandated private contributions to the SPR. Firms importing, refining, or domestically producing oil could be directed to supply specified amounts of oil to the SPR. The government could impose these costs on the firms to the extent that they were unable to pass the costs onto consumers, or subsidize them. Further, the oil or the market value of the oil could be guaranteed to each contributor in the event of reserve depletion.

These four options are not all mutually exclusive, nor would their implementation preclude continuing the current program. Each plan offers some advantages and disadvantages when compared to the others, and some combination of approaches may be appropriate to enable more flexible responses to a range of supply interruptions.

EVALUATION OF SPR FINANCING OPTIONS

Alternative ways of financing the SPR can be evaluated by several criteria. The most important of these concerns the distribution of the economic costs, benefits, and risks associated with the creation and depletion of the SPR.

Stockpiling oil entails a resource cost to society, as it displaces alternative economic activity and possibly more productive investment, and limits available funds for alternative social uses. Nevertheless, the reserve program benefits all sectors of the economy by reducing the adverse effects of oil supply interruptions. Drawing upon reserve oil during an interruption would bolster employment, contain inflation, and maintain the profit margins of firms. Finally, upon depletion, revenues would accrue to the owners of reserve oil, whether they be the federal government or private sector participants.

The costs for developing a reserve, however, can be shifted among various groups, depending on the financing method chosen. The total costs for alternative methods would differ slightly as a result of different institutional arrangements and economies of scale. In addition, Treasury debt financing would offer the lowest expected cost of all types of borrowing.

The bulk of SPR costs are inescapable, although they may be fully offset by the sales of SPR oil if the price of oil rises faster than the rate of return on an alternative investment, which is presumably the interest rate on Treasury bills. Thus, whoever finances the SPR assumes the risk that oil prices will not rise faster than the rate of interest. Under the current arrangement, in which oil is purchased through direct government expenditures, the taxpayers bear the risk. Plans that force firms to hold inventories or contribute oil to the SPR would transfer the burden to the oil industry, which, in turn, could pass on some of the costs to consumers. Since the abilities of individual firms to acquire and store oil vary, some firms would be placed at a competitive advantage. On the other hand, financing methods that rely on the speculative demand of individuals would shift the SPR costs from the taxpayers or firms to those individuals most willing to assume the risks in order to capture possible speculative gains. Private financing arrangements would allow the economy to benefit from the oil provided by the reserve during an interruption, but without requiring that taxpayers bear the risks regarding the rate of increase of oil prices. These financing options can also be measured against the following four criteria:

- The degree of federal control. To the extent that the SPR remains in the hands of the federal government, the option should preserve an appropriate level of federal control over reserve management.
- The budgetary effects. The option must have a predictable effect on the budget, and must provide SPR oil efficiently.
- The speed and level of acquisition. The option must allow for as rapid a buildup of the reserves as necessary.
- The producer nation response. The response of producer nations, which are opposed to the acquisition of SPR oil, must be assayed.

Both the public capitalization and debt financing plans would preserve maximum federal control over the SPR. Should the securities sold to the public to underwrite the SPR be denominated by the rate of oil price increase, it is possible that some SPR security holders might pressure for a depletion of the SPR wile its price is perceived to be at a maximum. Transferable titles, however, leading to an active secondary market for SPR securities, might diffuse this pressure. Mandating firms to contribute oil to the SPR would allow the government physical control of the oil. The contributors, having both an interest in the SPR and an active role in the oil market, might, however, be able to pressure the SPR administrator to their advantage. IPR options would reduce the federal government's control of the SPR. If firms were given control over their IPR reserves they might not choose to use them when the government sought to deplete the IPRs, opting to hoard the reserves in expectation of higher oil prices in the short term.

Despite these potential disadvantages, a combination of policies, with the associated mix of federal and private control might be appro-

priate. The federal government is unlikely to draw upon its centralized reserve in any situation short of a very severe supply interruption. During smaller disruptions, it might be desirable for the government to authorize the private sector to release IPR oil and allow firms
to bring their oil to the market as they see fit.

While the current program entails large expenditures, it does
allow the federal government to capture the revenues resulting from
reserve depletion. The budgetary effects of the four alternative financing plans would depend on both the structure of each plan and the degree to which it succeeds. Plans that call for off-budget financing
would make no significant demands on the federal budget, unless insufficient investment of private capital required supplemental federal expenditures to achieve desired fill rates. Public capitalization of the
SPR and plans that allow each firm to fill its IPR with oil held by
private individuals for speculative purposes fall within this group,
since the public would buy oil and receive the receipts of SPR depletion sales under these arrangements. Issuance of bonds yielding a
return determined by the rate of oil price increases could require
annual, but unpredictable, interest payments.

There is no guarantee that any of these speculation-based plans
would succeed in filling the SPR or IPR at the desired rate. If they
fell short, some backup system would be required to make up the
difference, presumably using public funds as in the current program.
These plans would have no budget effect for each barrel of oil that
they acquired, but could have a serious budgetary effect if they did
not succeed in attracting enough private capital to fill the SPR. This
effect would be equivalent to that of the current program for the same
quantity of oil, but without the benefit of advance planning.

Debt financing of the SPR at the market rate of interest would
have the budgetary effect of conventional debt financing by borrowing
money at prevailing market rates. Plans to provide incentives to
firms to hold extra inventories would affect the budget. Given the
uncertainties surrounding a firm's willingness to hold oil in response
to incentives and the extent of the incentives required, the budgetary
effect cannot be estimated with any degree of reliability. The budgetary effect of mandating oil contributions to the SPR would depend
largely on the degree to which the government subsidized the contributions. If, for example, the government provided the carrying charges
of the contributed oil, the short-term budget effect would be similar
to that of debt financing. Plans to decree that firms hold an IPR would
have no direct budgetary effect, but might well create equity problems
within the refining industry.

Those SPR financing plans that are based on speculative demand
for oil, public capitalization of the SPR, debt financing of the SPR
using securities denominated by the rate of oil price increase, and

plans for speculative purchase of firms' IPRs would not guarantee speedy completion of a strategic reserve. Plans to decree the existence of an IPR or to offer incentives for its creation might face difficulties in securing compliance. Inventories might be manipulated to present the illusion of compliance, or tank and pipeline bottoms (inventories that cannot be used) might be depicted as IPR oil. Mandating contributions would likely require noncompliance penalties but these might be easier to monitor. Debt financing of the SPR at market rates of interest, however, would provide enough financing to secure reserve oil if it is decided to buy it expeditiously.

None of the financing options would conflict with the goal of creating new storage capacity efficiently. New capacity can be drawn from three sources—new salt dome facilities, newly constructed aboveground storage, and renovation of old facilities. Salt dome storage capacity currently costs only $2 to $3 per barrel, compared to $12 to $16 per barrel for new aboveground steel tanks. The costs of renovation vary, depending on the specific characteristics of the existing facilities, the availability of which is unclear. Since salt dome storage capacity, the least expensive type, requires long lead times, an aggressive oil acquisition strategy would likely entail the higher costs of new, aboveground storage capacity.

Although producer nation response to stockpiling is a central issue in acquiring SPR oil, the possible responses by producers would not vary significantly among the options considered here. Both governments and firms offer advantages and disadvantages as SPR procurers. Producers could easily react to open government purchases with political threats of production cutbacks. Political considerations move in two directions, however, as the producer nations often require certain U.S. goods or services. Given their regular pattern of crude purchases, firms could possibly purchase oil that could be diverted to SPR uses without detection. Yet, the increasingly common "destination" contracts between OPEC nations and Western consumers, in which crude destinations are stipulated, would be abrogated if such diversions occurred. Firms might be reluctant to endanger their relationships with OPEC producers for the sake of the reserve. Therefore, none of the options considered in this chapter present a clear advantage in reducing the chance of adverse producer nation responses.

LEGISLATION TO REDUCE FEDERAL SPR COSTS

Recently, much congressional attention has focused on methods to reduce the federal expenditures for the SPR. Two recently introduced bills attempt to do this by shifting all or some of the costs to the private sector.

On March 4, 1981, Congressman Gramm introduced the Private Equity Petroleum Reserve Act (PEPRA), H.R. 2304, which would amend the Energy Policy and Conservation Act to finance the SPR through speculative private investment. The bill authorizes the Secretary of Energy to issue ten-year negotiable certificates, denominated in barrels of oil. It sets forth a number of terms, including the pricing mechanisms and exclusion from price controls and provides some of the details necessary to implement a public capitalization plan. Some problems remain unsolved; the price control exemption might be needed to attract investors, but could also commit the government to allow prices to rise during future interruptions. Since the bill establishes no restrictions or limitations on the size of the allowable investment, the capital market effects remain uncertain. Finally, the bill allows the immediate sale of oil currently in the reserve, which, if sold, would reduce the future flexibility of the SPR administrator.

On March 12, 1981, Senator Kassebaum introduced S. 707, the Strategic Petroleum Reserve Amendments of 1981. The bill mandates that each importer of more than 75,000 barrels of crude oil per day contribute five days of imports to the SPR annually. The government would pay each contributor an annual fee of 10 percent of the purchase price for 11 years. In the event of an emergency drawdown, the contributors would receive either oil or a payment for the oil. Any such payment would be equal to the world market price at the time of distribution, less any fees already paid, with a maximum payment of the average world price prevailing during the three months preceding drawdown.

The bill remains unclear about whether the government or the firms would own the oil after 11 years, since it omits any reference to the treatment of the oil after this period. Transferring ownership to the government would place a serious burden on the contributors, since they would not only be providing oil, but a very low interest loan to the government as well. If, on the other hand, the oil reverted to the firms, the federal government would burden the firms with the cost of the oil, but the 10 percent annual fee would cover some of the carrying charges.

While the Kassebaum bill would burden the contributing firms with some of the financial risks of the SPR, it would not provide a decentralized, privately held reserve to complement the SPR. In addition, it would provide little incentive for firms to reduce their conventional inventories. By focusing on crude oil and ignoring petroleum products, the bill also would tend to subsidize foreign refineries.

RELATED ISSUES

In addition to the evaluating criteria discussed above, several related issues should be kept in mind when considering alternative financing options. In devising a new SPR security, the Congress would determine where in the capital market it would compete for investor attention. SPR securities tied to the rate of oil price appreciation would be competitive with other "inflationary hedges," such as gold, real estate, or other commodity futures. Securities tied to the market rate of interest would be indistinguishable from other standard government securities, such as Treasury notes, bonds, and bills. The minimum investment required would also affect the position of SPR securities in capital markets. If SPR securities were issued in small denominations, they might compete with savings accounts, the predominant source of mortgage funds. On the other hand, high minimum investment requirements might preclude some classes of investors from buying the securities.

The primary advantage of moving the SPR program from the Department of Energy to a separate administrative body is to increase its independence. This independence might assist in filling the reserve through improving investor confidence in the reserve's management and financial integrity. Some measure of freedom from Civil Service personnel selection procedures, as is afforded the Synthetic Fuels Corporation, might help attract expert management from the financial community. On the other hand, this independence could diminish congressional control of the SPR program. Moreover, SPR purchases have to some degree been coordinated with other major consuming nations to reduce pressure on spot markets for oil. This coordination might suffer if the SPR is administered by an independent entity.

DEPLETION AND TERMINATION OF STOCKPILES

Private financing plans involving SPR securities based on oil price appreciation would create a set of claims upon either depletion, as part of a response to an emergency, or termination determining that the SPR was no longer needed. Both would involve the transfer of billions of dollars. The issues involved in depletion or termination of the reserve could have a dramatic impact on the U.S. economy.

Emergency depletion of the SPR will occur in a recessionary economic environment, presumably amid great uncertainty and conflict among income claimants. There might be a strong temptation to control oil prices in such an atmosphere. In the next several years, a U.S. shortfall of 1 million barrels per day could result in a sudden price increase of $20 per barrel. Moreover, if the disruption that catalyzed sudden oil price ratchets was clearly temporary, short-term oil price controls might preclude the large income transfers that distort the economy and reduce purchasing power dramatically.

Yet, the possibility of price controls on SPR sales would probably dissuade all potential purchasers of SPR securities. Controls and private financing might be compatible if SPR investors were given contractual guarantees of the equivalent of the world market price of oil upon depletion, rather than being given receipts of SPR oil sales per se.

Guaranteeing the equivalent of the world oil price raises the issue of the reference price for SPR sales. If SPR oil is sold by auction, payment to SPR investors could be calculated by prorating receipts. If SPR oil is sold at a price other than that established by this type of competitive process, some reference price would be required. The average price of U.S. imports during the week of transaction could provide a reference. Spot prices could also be employed. The choice of a reference price would affect the demand for SPR securities, given the characteristic pattern of higher spot prices during disruptions.

The treatment of any SPR security under the U.S. tax code must be established before any public offering. The attractiveness of an investment depends heavily on its tax treatment. For example, the demand for SPR securities would reflect whether deferred interest payments would be subject to a capital gains tax or taxed as ordinary income. In addition to altering the attractiveness or price of a SPR security, such decisions would affect the level of federal subsidy through tax expenditures.

The termination policy is a critical factor in public acceptance of stockpiling. If the SPR is not depleted during the term of SPR securities, these securities would have to be retired or rolled over. It might also happen that the SPR was never depleted, and the reserve would have to be terminated.

Under the SPR certificate plan, termination would involve the sale of the SPR oil and compensation to certificate holders, determined by the sale price of the reserve. This price would be influenced by the rate at which the SPR was depleted, the slower the depletion, the smaller the depressing impact on oil prices. Moreover, investors might eventually be wary of SPR certificates if the federal government retained the power to determine when the SPR would be terminated. A maturation date might have to be assigned to SPR certificates to provide a measure of certainty on this score. Maintaining the SPR beyond the maturation date would require marketing all SPR certificates again. If demand for these certificates was insufficient to maintain the reserve at the desired level, some new source of financing would be required, or a smaller SPR accepted.

Like SPR certificates, SPR bonds might require some maturation date. Such a termination date would allow for an examination of the need for the reserve after that period of time. A decision to maintain the SPR after its securities mature would require rolling over SPR bonds. It is unclear that demand would exist for these bonds if they were denominated by the rate of oil price appreciation. Continuing the SPR at that time might require additional federal expenditures. If SPR bonds were denominated by the market rate of interest, it would likely be fairly simple to refinance the reserve.

In the cases in which an Independent Petroleum Reserve is mandated, created by incentives, or contributions decreed, terminating the reserve would require eliminating the mandate or eliminating the incentive. Unlike under SPR certificates or bonds, which would be retired annually, the elimination of specific requirements or incentives could free up the entire Independent Petroleum Reserve at once. This would add substantially to oil market volatility. Under the evidence plan, termination would leave Independent Petroleum Reserve titleholders with oil rather than receipts. Since firms would no longer need to find individuals to hold speculative oil, firms would be forced to find new buyers for their oil and create some temporary market instability. The problems of a disorderly market could be reduced by a gradual phasedown of the Independent Petroleum Reserve storage requirements or incentives.

CONCLUSION

In the decades of the 1980s and 1990s, world oil supply is most likely to be less than the worldwide demand for oil. This perennial gap between demand and supply will force a continuous increase in world oil prices. The contributors in this volume agree that the world oil market will tighten considerably over the next 20 years. However, there is a wide range of expectations among experts regarding availability of supplies and price. For example, current forecasts of OPEC production range from 17 million barrels per day to 36 million barrels per day. The domestic political and economic pressures within oil-producing countries have induced implementation of policies to produce less oil than OPEC countries are capable of producing. Consequently, conservation of oil in the United States will lead to reduced production by foreign suppliers, but not to substantially reduced prices. The more pessimistic forecasts are based on the assumption that existing oil resources will be depleted, especially in the Soviet Union, and that oil-producing countries will continue to increase domestic consumption.

There probably will be repeated and unexpected disruptions of foreign oil supplies to the United States throughout the 1980s and the following decade. In concert with the long-term trend toward higher oil prices, these unexpected interruptions of oil production and delivery would cause additional price hikes and product shortfalls in the United States. As one contributor has estimated, there is an 80 percent probability of a 2 million to 3 million barrel per day shortfall in Persian Gulf oil during the next decade. Most analyses in this book focus on the Middle East, where supplies could be interrupted for a variety of reasons. The reasons range from domestic problems within the individual nations associated with rapid economic development that

301

has tended to destabilize these more traditional societies to conflicts between the nations in the Middle East. An external factor compounding the problem is the conflict between the United States and the Soviet Union over any change in the status quo within the region. The Soviet invasion of Afghanistan is a good example. These and other factors, individually or in concert, result in instability that could result in a major disruption of Middle East oil supplies. If the disruption existed for an extended period of time, it would have a devastating impact on the economy of the United States.

Foreign oil supply problems are not limited to the Middle East. For example, Indonesia may end its exports by 1990. Any shortfall in this oil product will be difficult to replace because of its high quality and low sulphur content. Historically, this source of high grade oil has been crucial in areas of the United States with severe air quality problems. For example, Indonesia now supplies 20 percent of California's oil needs and most of this oil is used for the generation of electricity in power plants. The consequence of a static Indonesian production rate and any increasing domestic demand, which would make this high grade oil less available, would be more severe and disruptive if accompanied by a shortfall in Persian Gulf oil.

It is clear that domestic oil supplies are inadequate to meet total United States demand and even some of that supply is vulnerable. For example, a major disruption of the Alaskan pipeline or its associated port facilities could cut off 40 percent of the oil supply to West Coast states for up to two years. A break in the pipeline itself could be fixed in hours or days and a broken pumping station in days or weeks. However, a catastrophe at the Port of Valdez caused by fire, earthquake, or military action could take months or even years to repair. The western United States could also lose some Alaskan oil if it were needed to fulfill United States international commitments to Japan or other allies during a major worldwide oil shortage.

The world oil market is also much less flexible today than it was during the 1970s. As a consequence, even small disruptions in the future could have more serious effects. During earlier oil shortages, for instance in 1973-74, oil companies traded crude extensively, which minimized the impact of the shortfall. During the 1978-79 shortfall and even more recently, oil-producing countries have taken much greater control of their crude supplies and have placed restrictions on how much oil can go to what destination and under what conditions. Several nations have even forced oil companies to accept various kinds of political restrictions in contracts. The contributors in this volume generally agree that as a consequence of new contractual arrangements dictated by oil-producing countries, the world market will be far less able to adjust to any future shortfalls.

The physical characteristics of oil have also changed in recent years with the average crude oil sold on world markets becoming increasingly heavy. Unfortunately for consumers in the United States, the price differential between heavy and light crudes will increase as lighter crudes become more difficult to find. In addition, the new supplies from Alaska, Mexico, and other sources are likely to be relatively heavy. This will increase the current trend toward a glut of heavy residual oil in the western United States.

On the other hand, and fortunately for the United States, natural gas will be increasingly available on the world markets. Both Mexico and Canada will be able to increase pipeline exports of natural gas to the United States and gas from more remote sources will be available as liquefied natural gas or possibly as methanol. Given the relatively clean burning characteristics of natural gas and methanol, they could be used to replace at least some of the rapidly diminishing light crude oil supplies.

While energy analysts in the United States confront the complexities of the problems of international oil supply and demand, which the contributors in this text have characterized as unpredictable and insecure in nature, they are facing a concurrent set of domestic problems. For instance, how does an energy analyst design and implement an information network that ensures that they have relevant and accurate information to manage an oil shortage? What kind of mechanism best allocates the scarce oil resources during a severe shortage in order to meet the public and private interests of the nation? In the longer run, what kind of policies should the national government adopt to reduce vulnerability to shortfalls in foreign oil supplies?

There is a general consensus among the contributors that current federal policies and programs to manage shortfalls are inadequate. The strategic petroleum reserve is being filled slowly, and there is no agreement about the circumstances under which it will be used. In addition, federal policy in 1973-74 was to control gasoline prices, allocate products, and make the states responsible for managing the effects of any shortages, such as long lines and fuel redistribution problems. Without changes now, these inadequate past policies are likely to be reimposed by Congress in the event of an emergency and, most of the contributors agree, they will not permit the United States to cope with the problems.

One conclusion derived from these readings is the need for United States policymakers to distinguish between long-term oil market supply shortages and short-term, unexpected disruptions. The long-term problem must be resolved by developing new oil supplies, investing in conservation, commercializing synthetic fuels, and deploying alternative technologies. However, these approaches will not protect the nation against unexpected cutoffs of oil supply, because dependence on oil will continue for at least the next 10 to 20 years. Measures to deal with such disruptions include stockpiling, development of surge production capacity, and emergency measures to reduce oil demand. The retrofitting of domestic oil refineries falls into both categories. It may be needed to adapt to increasingly heavy crude oil supplies, which also will decrease vulnerability to disruptions by increasing the flexibility of U.S. refineries to process more diverse crude oil products.

Most analysts recommend dealing with the long-run oil supply problem by allowing oil prices to rise. However, the general public sees price increases as the major problem. This conflict is central to the ongoing debate over the impact of decontrol of oil prices as a means to secure additional oil supply at a reasonable price. Whatever fuel allocation mechanisms are developed by the United States government, these mechanisms for managing unexpected disruptions must be in place before the emergency occurs. Otherwise, the measures will neither forestall panic nor be politically or practically implementable.

Private sector stockpiling should be used to supplement government stockpiling. Most experts agree that private companies can stockpile oil more efficiently than the government, albeit at a higher cost. Government, however, could perform the following functions. First, it could require private firms to stockpile and permit the companies to pass the costs of storage along to consumers. Tax or other incentives to stockpile could be enacted and private firms could be allowed to store oil in an unused Strategic Petroleum Reserve. Perhaps most importantly, the government could help establish a privately owned national storage corporation similar to the West German system. The contributors disagree on the issue of whether or not the decision to release private stocks should be made by the companies, an independent board, or elected officials. There seems to be a consensus that the companies involved were likely to be too cautious about releasing their stocks and thereby dilute the effectiveness of the program.

During a severe fuel shortage both market and nonmarket allocation mechanisms would have diverse impacts. For example, white market coupon rationing, an emergency gasoline tax, or general oil decontrol could end gas lines caused by oil supply disruptions. However, the reimposition of price controls during a serious disruption would lead to gasoline lines, which may or may not be reduced by measures such as odd/even rationing, carless days, or other requirements. Three alternatives were discussed by the contributors. First, coupon rationing, which may require imposing price controls and issuing what amounts to a second currency, could be established. Second, an emergency gasoline tax with a rapid rebate would be a similar fuel allocation mechanism, but it would be simpler to administer. The third alternative would be to decontrol all oil prices, collect windfall profits, lower taxes, and raise welfare and social security benefits as a rebate. The first two would cause the consumer price index and indexed payments to rise and would load the whole shortage onto gasoline, foregoing the opportunity to save jet fuel and heating fuel. If the oil shortage is prolonged, all options except decontrol would lead to misallocation of resources.

Finally, conservation to reduce U.S. dependency on foreign oil must be kept in perspective. Unfortunately, conservation alone is not adequate to manage major disruptions. Reducing oil demand will reduce internal pressures in oil-producing countries and reduce the probability of disruptions. However, conservation can actually exacerbate the problem. Because the least valuable uses of oil are conserved, shortfalls are even more painful when they do occur. Nevertheless, conservation is still desirable because it reduces the balance of payments deficit, and improves the negotiating position of the United States with other nations.

There are costs and benefits to every decision proposed by policymakers to gather and validate the necessary information to manage oil production and distribution, to design an allocation mechanism for implementation during a crisis, or to enact energy policies that reduce U.S. vulnerability to foreign oil supply disruptions. These cost/benefit issues will remain at the center of much of the domestic and foreign policy debate in the United States throughout this century. Indeed, energy decisions about who gets what, where, when, and how and who pays for it, whether for the good of the individual or the public, will probably remain the paramount issue as the United States and nations throughout the world address the critical issue of energy sources.

ABOUT THE EDITORS
AND CONTRIBUTORS

LARRY L. BERG holds a B.A. and M.A. in political science and history from the University of Iowa, Iowa City, and a Ph.D. from the University of California, Santa Barbara. Dr. Berg is an associate professor of political science at the University of Southern California and directs the USC's Institute of Politics and Government, and its Politics in Washington Program. As an authority on judicial elections, political corruption, and the initiative process, he has lectured extensively in the United States, Great Britain, and the Soviet Union. He is coauthor of three books: Corruption in the American Political System, with Harlan Hahn and John R. Schmidhauser; The Challenge of California, with Eugene Lee; and The Supreme Court and Congress: Conflict and Interaction, 1945-1968, with John R. Schmidhauser. His numerous articles have been published in journals such as Social Science Quarterly and Law and Policy Quarterly.

LAWRENCE M. BAIRD is presently an advisor to Commissioner Emilio E. Varanini III, California Energy Commission. Prior to this, Dr. Baird was principal consultant to the Assembly Office of Research. He received his doctorate from the University of Southern California in 1978. His publications include studies concerned with the environment of the southern California coastal zone, public administration in the 1970s, and policies for improving the California legislature.

Commissioner EMILIO E. VARANINI III is an engineering graduate of the U.S. Naval Academy and holds a law degree from the University of Pacific. He has been involved in international, national, and regional energy issues for the last 12 years as a Chief Consultant to the California Legislature, and as a Commission on the California Energy Commission. Currently, he directs one of the most comprehensive strategic energy planning and emergency preparedness efforts in the United States.

TOM E. BURNS is supervisor of the supply division of the economics department at Stanford Oil Company of California. He has spent 17 years in the oil industry, of which 8 were in Europe.

DAVID A. DEESE directs Harvard's Energy and Security Project and teaches in the Kennedy School of Government, where he serves as assistant to the director at the Center for Science and International Affairs. He has published books and articles on various aspects of the problems of radioactive waste and nuclear nonproliferation, and is coeditor with Joseph Nye of the book Energy and Security.

DENNIS EOFF is an economist and oil analyst for the California Energy Commission. He co-authored the commission's 1980 study Fuel Price and Supply Projections.

FEREIDUN FESHARAKI is a research fellow at the Resource Systems Institute of the East-West Center and adjunct associate professor of economics at the University of Hawaii at Manoa. He was formerly the energy advisor to the prime minister of Iran and a delegate to the OPEC Ministerial Conference. He has published many articles on oil, OPEC, and Iran.

ARTURO GANDARA has been on the staff of the Rand Corporation's Social Sciences Department for several years. His principal research interest is government regulation, and his publications include the Rand study Electric Utility Decision-Making and the Nuclear Option.

R. D. HALL, an engineer, serves on the Alberta Petroleum Marketing Commission, for which he was formerly executive director for economics and planning. He has also worked for the Canadian National Energy Board.

CARLYLE HYSTAD is a project manager at Sobotka and Co., Inc. He was formerly the director of the Office of Emergency Response Planning and the deputy director of the Strategic Petroleum Reserve at the U.S. Department of Energy.

SUSAN L. MISSNER is manager of Stanford University's International Energy Program and a research associate at the graduate school of business. She was formerly with Planning Research Corporation, where she was extensively involved in assessment of federal solar programs.

THOMAS L. NEFF is a principal research scientist and manager of the International Energy Studies Program at the M.I.T. Energy Laboratory. Dr. Neff has also served as an advisor to the U.S. Department of State, U.S. Arms Control and Disarmament Agency, and the Office of Science and Technology Policy. Among his publications are forthcoming books on solar energy and the international uranium market.

GARY J. PAGLIANO analyzes domestic and international energy policy issues for the Congressional Research Service, the research arm of the U.S. Congress. He was formerly a researcher for the Wednesday Group in the House of Representatives. His recent publications include "Mexican Oil and Gas Policy," and "The Oil Market Impact of the Iran/Iraq Conflict and its Potential Expansion."

GUY PAUKER is a senior staff member of the Rand Corporation and serves as the executive secretary of the Asia-Pacific Energy Studies Consultative Group, which meets under the auspices of the East-West Center. Formerly a professor at the University of California at Berkeley and a lecturer at numerous other institutions, he has written many books and articles on political science, Asia, and Indonesia.

ALICE M. RIVLIN is Director of the Congressional Budget Office. She received her M.A. and Ph.D. degrees from Radcliffe. She has served as senior staff economist and senior fellow to the Brookings Institute. Dr. Rivlin is a member of the American Economics Association.

DAVID RONFELDT has been on the staff of the Rand Corporation's Social Sciences Department for several years. He has served as the principal investigator on projects for the U.S. Departments of State and Energy, and has spent the last three years studying relations between the United States and Mexico. With Richard Nehring and Arturo Gandara, Ronfeldt co-authored the recent Rand publication Mexico's Petroleum and U.S. Policy.

JOHN J. SCHRANZ, JR. is currently a Senior Specialist in the Congressional Research Service, Library of Congress, working in energy resources policy. Prior to 1979, Mr. Schranz was at Resources for the Future, where he was a Senior Fellow in their Energy Division. He recently completed the coordination of a CRS study for the House Energy Committee on energy consumption. Schranz has written articles in the areas of mineral resources and energy.

WILLIAM B. TAYLOR JR. is the director of the Office of Oil Supply Security, Policy and Evaluation, U.S. Department of Energy. He has also worked for other DOE offices, the Congressional Budget Office, and Harvard University's Kennedy School of Government.

FRANK TUGWELL is deputy assistant administrator, responsible for energy and natural resource programs, at the U.S. Agency for International Development. On leave from Pomona College where he is an associate professor, Dr. Tugwell has also worked for the U.S. Congress' Office of Technology Assessment and the U.S. State Department's Office of Fuels and Energy. He has lived in Venezuela and is the author of The Politics of Oil in Venezuela.

STANSFIELD TURNER was Director of the CIA in the Carter administration. In 1975, Turner was appointed Commander-in-Chief of the Allied Forces in Southern Europe (NATO). He received his

B.S. from the U.S. Naval Academy and his M.A. from Oxford University where he was a Rhodes scholar. He also holds Hum.D., D.Sc., and LL.D. degrees.

ARLON TUSSING is a consultant, a professor at the University of Alaska's Institute of Social and Economic Research, and a member of the Economic Advisory Board of the U.S. Department of Commerce. He was formerly the chief economist for the U.S. Senate Committee on Interior and Insular Affairs. He has published numerous papers on West Coast petroleum problems.